The Measure of Times Past

Donald J. Wilcox

The Measure of Times Past

Pre-Newtonian Chronologies and the Rhetoric of Relative Time

The University of Chicago Press
Chicago and London

Donald J. Wilcox is professor of history at the
University of New Hampshire.

The University of Chicago Press, Chicago 60637
The University of Chicago Press, Ltd., London

© 1987 by The University of Chicago
All rights reserved. Published 1987
Printed in the United States of America

96 95 94 93 92 91 90 89 88 87   54321

Library of Congress Cataloging-in-Publication Data

Wilcox, Donald J.
    The measure of times past.

    Bibliography: p.
    Includes index.
    1. Chronology, Historical.   2. Historiography.
I. Title.
D11.W59   1987   902'.02   87-1669
ISBN 0-226-89721-4
ISBN 0-226-89722-2 (pbk.)

To
J. R. G.
Forsan et haec olim meminisse juvabit

Contents

Acknowledgments

I wish to thank the National Endowment for Humanities and the American Council of Learned Societies for their generous support, which enabled me to devote a full sabbatical year to the final stages of research and writing. I am also grateful to the Rockefeller Foundation for providing the magnificent environment of the Cultural Center at Bellagio for a most productive period of rewriting and finishing the final drafts. Finally I appreciate the support of the Wheeler Fund of the History Department at the University of New Hampshire for its support of my research travel.

Several people have read and criticized parts of the manuscript, including Alan Bernstein, Mark Phillips, William Bouwsma, John Marquand, and David Walters. I appreciate their interest and their suggestions. In addition, parts of the argument have been presented to history department seminars at the universities of New Hampshire, California at Berkeley, California at San Diego, Oregon, and Washington. Reactions to those seminars were most helpful in refining and formulating the final manuscript. Finally, my students at the University of New Hampshire have been among my most penetrating and helpful critics as I have exposed these ideas in my seminars and courses over the past several years. I am profoundly grateful to them, and the book owes much to their reactions.

1

Introduction

"Mr. Palomar is standing on the shore, looking at a wave. Not that he is lost in contemplation of the waves. He is not lost, because he is quite aware of what he is doing: he wants to look at a wave and he is looking at it. He is not contemplating, because for contemplation you need the right temperament, the right mood, and the right combination of exterior circumstances; and though Mr. Palomar has nothing against contemplation in principle, none of these three conditions applies to him. Finally, it is not 'the waves' that he means to look at, but just one individual wave: in his desire to avoid vague sensations, he establishes for his every action a limited and precise object."[1] As the wave breaks up into other waves, Mr. Palomar does not lose heart. He changes viewpoint, enlarges and contracts his field of vision, and perseveres in the confidence that, should he succeed in seeing the wave as a single object, "it could perhaps be the key to mastering the world's complexity by reducing it to its simplest mechanism" (p. 6). But the waves prove recalcitrant; they are too shifting, too complex, and too interrelated. Abandoning his quest, "Mr. Palomar goes off along the beach, tense and nervous as when he came, and even more unsure about everything" (p. 8).

Italo Calvino has drawn his protagonist as an exemplar of twentieth-century humanity seeking the certainty of concrete and objective truth in a world where particular events with simple locations in space and time are no longer seen as adequate expressions of reality. At first it might seem that Mr. Palomar in his anxious pursuit of a fixed and complete knowledge of his world has not heard of those changes in twentieth-century science that call into question the possibility of such knowledge. But such is not the case. Though he once felt he could construct a single conceptual model whose geometrical precision would correspond to the data of experience, the difficulties of implementing such a goal had forced him to change his approach. "Now he needed a great variety of models, whose elements could be combined in order to arrive at the one that would best fit reality, a reality that, for its own part, was always made up of many different realities, in time and in space" (p. 110).

Mr. Palomar's problem is not ignorance but anxiety and doubt. He does not know how to grasp experience in this new way, how to incorporate his own life into that multifarious world of nature which scientific theory has created, or even how to preserve his own identity when there is no fixed point of certainty from which knowledge can grow. "The swimming ego of Mr. Palomar is immersed in a disembodied world, intersections of force fields, vectorial diagrams, bands of position lines that converge, diverge, break up. But inside him there remains one point in which everything exists in another way, like a lump, like a clot, like a blockage: the sensation that you are here but could not be here, in a world that could not be but is" (p. 17).

Most troubling of all to Mr. Palomar are the uncertainties of space and time. Trying to look at the stars, he cannot find the right place from which to make the observations; he cannot manage to look at the chart and the stars at the same time, and his contortions in the effort are "like the convulsions of a madman" (p. 48). On a visit to the reptile house he contemplates the slow and apparently aimless movements of the iguanas, wondering whether they exist in another kind of time from his.

> In what time are they immersed? In that of the species, removed
> from the course of the hours that race from the birth to the death
> of the individual? Or in the time of geological eras, which shifts
> continents and solidifies the crust of emerged lands? Or in the
> slow cooling of the rays of the sun? The thought of a time outside
> our experience is intolerable. Mr. Palomar hurries to leave the
> reptile house, which can be visited only now and then and in
> haste. (pp. 87–88)

Mr. Palomar's flight from the reptile house and the multiple times it conjures up is a true metaphor of the twentieth century, for we have witnessed more profound changes in basic constructs of space and time than any period since the seventeenth century, when the Scientific Revolution laid down the notions of space and time that bounded the Newtonian world. Contemporary thinkers and artists in fields as diverse as physics, painting, poetry, psychology, and philosophy have challenged the sense of time and space as objective, continuous, all-embracing, and absolute that they inherited from the Newtonian revolution. In place of this sense of absolute time and space they have substituted a new view, one which sees space and time as subjective, fragmented, relative, overlapping, and disconnected.[2]

Einstein and Mach, Mallarmé and Apollinaire, Proust and Joyce, Picasso and Boccioni, Bergson and Freud all worked toward this reorientation. In some cases the participants in the revolution were acutely conscious of achievements in other fields and looked on their work as part of a large change fraught with conceptual significance. The Futurists saw in the early cinema a realization of their sense of simultaneity (Kern, pp. 71–72), and Proust sensed in the music of Franck a representation of his own sense of time. In other cases writers in one field disavowed the metaphysical or universal application of their ideas. Most significantly, Einstein was skeptical of metaphysics in general and never acknowledged that his special theory of relativity contradicted the bases of classical science. He sought to integrate his own modifications into the framework of classical Newtonian science, just as Copernicus in the sixteenth century had tried to fit his solar system into the mechanics of Aristotelian physics.

Unsophisticated attempts to extend the implications of relativity theory into ethics and social analysis in the twenties and thirties lent weight to Einstein's disclaimer. Their failure reinforced the validity of Newtonian science for real life, relegating the human application of the new time and space to modern art and literature. But the separation between art and life is not complete. In some cases practical achievements have depended directly on artistic representation of the new time and space.

One of the most striking examples of the connection between practicality and art lies in the development of camouflage in World War I. Picasso, when he saw the first camouflaged truck moving down the streets of Paris, observed to Gertrude Stein that he and the other Cubists had created that form. He saw instantly the affinity. Both camouflage and Cubism overturned the distinction between background and foreground that had dominated Western visual arts since the Renaissance, a distinction deeply intertwined with the Scientific Revolution and with Newtonian time and space. But Picasso's comment was not simply a brilliant artistic insight. The affinities he saw were not accidental. The inventor of camouflage, Giaurand de Scévola, a telephone operator for an artillery unit, claimed that he had indeed learned the technique of camouflage from the Cubists, and furthermore he used Cubist painters in the early stages of design. "In order to totally deform objects, I employed the means Cubists used to represent them—later this permitted me, without giving reasons, to hire in my [camouflage] section some painters, who, because of their very special vision, had an aptitude for denaturing any kind of form whatsoever" (in Kern, pp. 302–4).

The new time of the twentieth century poses special problems for historians. Not only are they particularly concerned with events in time, but the practice of historical research in the modern era is closely linked, both conceptually and historically, to Newtonian time. The continuous and universal qualities of Newton's time and space have made it possible for historians—as well as the natural scientists—to view the basic components of reality not as processes or organic wholes but as a series of discrete events that can be placed on a single time line and at a single point in space. Large historical trends, major events, overarching generalizations, and broad issues of causality and direction remained as important as they had been in the works of earlier historians, but with the dominance of absolute time and space they became secondary realities, agglomerations of more basic units of historical study.

Working historians seldom have occasion to think about such metaphysical issues as the implications of absolute time and most often prefer to dismiss them as irrelevant to the practice of history. After all, not all conceptual implications contained in an idea need find expression in reality. The existence of a tendency in Newtonian time to reduce historical events to simple, primary facts and discourage the perception of realities that could not be so reduced might be overcome in many ways. Since historians more often than not operate without even a conscious awareness of such a tendency, it has been commonly ignored by the historical profession. Many scientists even argue that these features of absolute time and space have no particular relevance in natural science. Newton had many theological interests which were not connected with his purely scientific work. Especially in the General Scholium to his *Principles*, which he added in later editions, he introduced a number of metaphysical and theological concepts which give expression to his religious needs but which were not essential to his laws of motion. Nevertheless Newton's laws of motion are dependent on absolute time and space, and those concepts occur not in the General Scholium but at the beginning of the *Principles*, where he defined the concepts he considered essential to his laws.[3]

By the same token modern historians operate under the assumption of an absolute and continuous time line, even if they are not always immediately aware of it, and this assumption colors their sense of what history is about. Marc Bloch, certainly one of the greatest and most sophisticated historians of this century, defined history as the science of men in time. In defining this time and distinguishing the historian's approach from that of the natural scientist, Bloch presented clearly the notion of individual historical events which take on their meaning by being located as points within a general and continuous time.[4]

> Historical time is a concrete and living reality with an irreversible onward rush. It is the very plasma in which events are immersed, and the field within which they become intelligible. . . . No historian would be satisfied to state that Caesar devoted eight years to the conquest of Gaul, or that it took fifteen years for Luther to change from the orthodox novice of Erfurt into the reformer of Wittenberg. It is of far greater importance to him to assign the conquest of Gaul its exact chronological place amid the vicissitudes of European societies; and without in the least denying the eternal aspect of such spiritual crises as Brother Martin's, he will feel that he has given a true picture of it only when he has plotted its precise moment upon the life charts of both the man who was its hero and the civilization which was its climate.

This sense that the most elemental act of the historian is the location of single events in time should not be confused with the issue of historical objectivity. The absolute time line was indeed an essential support for many nineteenth- and early twentieth-century historians who felt that history consisted of irreducible and indisputable concrete facts that were immune to all interpretive distortion. Lord Acton, for instance, when he sought to make the *Cambridge Modern History* a work that would transcend all bias and personal interpretation, conceived that work in absolute space rather than the relative geography of nations. "Contributors will understand that we are established not under the Meridian of Greenwich, but in Long. 30 degrees W.; that our Waterloo must be one that satisfies French and English, Germans and Dutch alike; that nobody can tell, without examining the list of authors, where the Bishop of Oxford laid down the pen, and whether Fairbairn or Gasquet, Liebermann or Harrison took it up."[5]

Acton's insistence on objectivity certainly depended on the concept of ideally isolated events unequivocally existing in time and space, but the absolute time line and the spatial coordinates that accompany the concept permeate most modern historical thinking, even when the focus is on generalization, on universal history, and upon secular historical trends. In fact it is often the very minuteness and number of the individual events that have led historians to insist on the importance of generalizations. As Henri Pirenne described the historian's metier, "The subject of the historians' study is the development of human societies in space and time. This development is the result of billions of individual actions. But in so far as they are purely individual, these actions do not belong to the domain of history, which has to take account of them only as they are related to collective movements, or in the measure to which they have influenced the collectivity."[6]

The task of reducing complex processes to the simple events that constitute them often produces in historians the same sense of confusion and frustration that Mr. Palomar experiences as he tries isolate the waves on the seashore. Listen to the great intellectual historian Carl Becker, as he addresses the question of what constitutes a fact.

> First then, what is the historical fact? Let us take a simple fact, as simple as the historian often deals with, viz.: "In the year 49 B.C. Caesar crossed the Rubicon." A familiar fact this is, known to all, and obviously of some importance, since it is mentioned in every history of the great Caesar. But is this fact as simple as it sounds? Has it the clear, persistent outline which we commonly attribute to simple historical facts? When we say that Caesar crossed the Rubicon we do not of course mean that Caesar crossed it alone, but with his army. The Rubicon is a small river, and I don't know how long it took Caesar's army to cross it; but the crossing must surely have been accompanied by many acts and many words and many thoughts of many men. That is to say, a thousand and one lesser "facts" went to make up the one simple fact that Caesar crossed the Rubicon; and if we had someone, say James Joyce, to know and relate all these facts, it would no doubt require a book of 794 pages to present this one fact that Caesar crossed the Rubicon. Thus the simple fact turns out to be not a simple fact at all. It is the statement that is simple—a simple generalization of a thousand and one facts. (in Meyerhoff, pp. 121–22)

Two features of this passage are especially noteworthy: the reference to Joyce and the use of the A.D./B.C. system to characterize a simple fact. Joyce, together with Proust and Mann, was among the novelists who in the first half of this century worked to devise a narrative style that would present reality outside of the traditional constructs of space and time. Their works represent a new approach to experience, an attempt to see life in multiple times and as a group of related processes that cannot be reduced to simple elements. They looked at human life organically, as part of a totality of experience. Becker's awareness that he cannot reduce the crossing of the Rubicon to its component elements, like Mr. Palomar's

failure with the waves, is part of the consciousness that produced and accepted this major revolution in twentieth-century writing.

If the reference to Joyce shows Becker's acceptance of twentieth-century
trends, the type of fact he chose as an example demonstrates his commitment to older notions of space and time. Just as Lord Acton used a numerical measurement, longitude, to illustrate the complete objectivity he
wanted for Waterloo, Becker chose as a typical fact one to which he could
give a precise numerical location in time. The number he used to locate
the crossing of the Rubicon, 49 B.C., is part of the dating system in universal use, at least in the Western world. He could have dated the crossing by
telling us Caesar's age at the time, or he might have told us what office
Caesar held in the republic when the event occurred. He might even have
dated it in reference to some other event, like the Roman civil wars or the
consulship of Cicero. Contemporary historians do use such relative dates
in their writing, but only for literary effect or to bring out a specific theme.
When we wish to present an event as a naked fact, the surest way to do so
is to give it its own date.

This dating system is a truly remarkable instrument and worthy to be
the symbol of objectivity and certainty for historical facts. It can perform
the most minute and immediate quotidian tasks—organize our check-
books, fix our entry into school or our retirement from the work force, and
calculate fiscal and legal responsibilities. The same dating system can locate such distant events as the battle of Marathon, the period when people
first engaged in agriculture, and even the time when life on earth began.
Precision certainly diminishes as the times retreat into the distant past,
but the same series of numbers, extending indefinitely forward and backward, serves to locate vastly dissimilar events within a single, unambiguous,
and absolute time frame. Our comfortable familiarity with this dating system makes it seem a natural means of expressing temporal relationships.
Certainly by quantifying the temporal interval between events it makes
them more susceptible to analysis. As one student of ancient chronology
has remarked, "Time is the proper dimension of history. A fact is historical
when it has to be defined not only in space but also in time. A fact is
placed in the fourth dimension, that of time, by measuring its distance
from the present. Chronology, an auxiliary of history, enables us to state
this time-interval between a historical fact and us by converting the
chronological indications of our sources into units of our own time
reckoning."[7]

The B.C./A.D. dating system displays all of the features of Newtonian
time. Indefinitely extended forward and backwards from an arbitrary

point, it is truly universal in application and seems to carry with it no thematic or interpretive weight. It can date events which have no intrinsic relationship to one another and quantify their temporal distance precisely and unambiguously. It is a continuous, universal, and endless time line with no external reference, a chronology perfectly adapted to the needs of Newtonian time. Though the system might seem tied to the birth of Christ, its inventor considered that reference point to be purely arbitrary. He was fully aware of the discrepancy between the actual birth date of Christ and the conventional date used in the B.C./A.D. system.

The bonds between the modern dating system and Newtonian time are not simply conceptual; they are historical. The system was devised and implemented during the same century in which the ideas were formulated that led to Newton's great work. A Jesuit scholar named Domenicus Petavius published the key work setting out the B.C./A.D. system in 1627, ten years before Descartes proposed in his *Discourse on Method* some of the essential notions behind the new scientific methodology. Petavius' system came slowly into use during the seventeenth century, with many chronologers preferring an older system, named the Julian Period. Newton himself, in a chronological work published in 1728, was among the first scholars to use the new system exclusively, and it became the standard dating system in the West only during the eighteenth century. Not only did Newton find it congenial, but those who rejected the notion of absolute time, such as Vico, were also among the few critics of the system in the eighteenth century.

The B.C./A.D. dating system allows us to quantify the relationship between any two events with a precision no ancient historian could imagine and to express this precision with a single number. Furthermore, we can also use numbers to identify general characteristics of an era that might not so easily be traceable to a single event. Historians can refer to the "1930s" whether they are discussing the economic history of the Depression, the political history of the New Deal, or the international history of fascism. These numerical terms are important to our claims to historical objectivity, for they provide the units by means of which different views of the past can be discussed. One historian, considering the dominant trait of the early nineteenth century to be industrialization, might refer to the period as the Age of the Spinning Jenny; a second, having in mind the romantics' interest in nature and the emotions, might call it the Age of Wordsworth; while still a third, calling attention to the spread of revolutionary ideas and institutions, might call it the Age of Napoleon. But they would all agree on what events they could exclude from the period, since all understand the meaning of the term "nineteenth century." It has a quantitative sense which seems to transcend interpretation and bias.

Yet the use of centuries barely antedates the B.C./A.D. dating system. Although Moslem historians had long organized events by centuries, the first major work in the Western world to mark periods in such an abstract form was the *Magdeburg Centuries*, an ecclesiastical history written by a group of Lutheran scholars in 1559. Though it appeared during the Reformation, the concept of centuries did not come into widespread use before Newton's time. Not until after the seventeenth century did most literate contemporaries identify the epoch in which they lived as their "century," and only in the course of the Enlightenment did the term take on the epochal significance we now attach to it.[8] Scholars now enjoy the luxury of complaining that centuries are meaningless as indicators of major changes, but they could not even decry the use of such epochs if they had no sense that abstract quantities could contain and characterize a group of events occurring within them.

The dating systems in use before Newton were not absolute and did not contain the implications about absolute time that characterize the B.C./A.D. system. Pre-Newtonian time had no conceptual grid to give universal applicability to its numbers. Dates in that time were tied to specific themes, events, and moral lessons, and they gave a meaning and shape of their own to the events they dated. Without the conceptual grid the fundamental sequence of measurement and judgment was radically different from ours. Whatever our attitude toward the function of history and the possibility of objectivity, we assume that a historian answers the question of when an event occurred before determining its meaning; judgment follows measurement. Historians writing before Newton reversed this order. Events created their own time frames. Before locating an event in time, the historian had to make judgments about its meaning and its thematic relation to other events. New insights, syntheses, or major events, such as the rise of Rome, the coming of Christianity, the development of modern nations, or the revival of culture in the Renaissance all created their own time frames.

For historians before Newton the time frame did not include a group of events; a group of events contained a time frame. This perspective led them to use a variety of relative dating systems, none of which had an absolute temporal significance apart from the group of events that gave it its meaning. Since we use an absolute time line as a basis for our synthetic understanding of the past, modern scholars have often viewed the relative time frames of early historians with condescension, calling relative time simple and primitive. One of the foremost authorities on chronology in the twentieth century has observed, "The simplest and most ancient method of dating is the relative time-reference which does not require any chronological devices. . . . [In primitive tribes,] except for savants, men have little interest in absolute time notations; they use, instead, relative time-references. Primitive peoples usually do not know how old a man is, but only who is the oldest in the group" (Bickerman, p. 62). In fact relative dating in so-called primitive tribes betrays only the most superficial similarity to the chronologies of Herodotus, Thucydides, Polybius, Augustine, Machiavelli, or the other historians who wrote before Newton. Nevertheless the absence of absolute dating systems from their works has helped create some of the common misconceptions about their sense of time, particularly with the classical historians. Without an absolute dating system they could more easily be accused of lacking a true sense of time and of viewing events from a framework in which only unchanging concepts had any reality.[9]

If the philosophers, writers, and artists of the twentieth century are correct, then there are aspects of experience which cannot be treated as simple concrete events existing as points on a time line, even though they have unmistakable temporal dimensions. Historians are slowly coming to acknowledge the importance of such realities. In one of the seminal works of contemporary historiography, Fernand Braudel has sought to tell the story of the Mediterranean world during the late sixteenth century. In the introduction to that work, Braudel, writing in 1946 as a member of the Annales School that had been founded in part by Marc Bloch, described a variety of times that made up the parts of his book.[10]

> The first part [of the book] is devoted to a history whose passage is almost imperceptible, that of man in his relationship to the environment, a history in which all change is slow, a history of constant repetition, ever-recurring cycles. . . . On a different level from the first there can be distinguished another history, this time with slow but perceptible rhythms. . . . Lastly, the third part gives a hearing to traditional history—history, one might say, on the scale not of man, but of individual men, what Paul Lacombe and François Simiand called *l'histoire événementielle*, that, is the history of events: surface disturbances, crests of foam that the tides of history carry on their strong backs.

In this passage Braudel pointed to a time which existed between the timeless realities of rise and fall, seasonal change, and the constants of human nature on the one hand, and the hurried, individual events that could be dated by the modern dating system on the other. He called this intermediate time the time of the *longue durée*. In an article published a decade later in *Annales*, Braudel explained more clearly that the longue durée characterized certain special kinds of events that could not be comprehended accurately in l'histoire événementielle, especially events in economic, social, and intellectual history.[11]

> The recent break with the traditional forms of nineteenth-century history has not meant a complete break with the short time span. It has worked, as we know, in favor of economic and social history, and against the interests of political history. . . . There has been an alteration in traditional historical time. A day, a year once seemed useful gauges. Time, after all, was made up of an accumulation of days. But a price curve, a demographic progression, the movement of wages, the variations in interest rates, the study (as yet more dreamed of than achieved) of productivity, a rigorous analysis of money supply all demand much wider terms of reference.

In addition to economic and social trends, Braudel also pointed to cultural affairs as a realm where changes occurred in the time of the longue durée, citing the studies of mentality among French historians in the past generation, including such works as Lucien Febvre's investigation of attitudes towards religion in the sixteenth century and Alphonse Dupront's examination of the idea of the crusade in late medieval Europe (*On History*, p. 32).

Braudel knew how threatening the hypothesis of another time scheme would be to his colleagues in the historical profession. In his defense of the longue durée he assured his readers that it would not replace ordinary historical time, even while he insisted that historical time did not comprehend certain important realities of life.

> Among the different kinds of historical time, the longue durée often seems a troublesome character, full of complications, and all too frequently lacking in any sort of organization. To give it a place in the heart of our profession would entail more than a routine expansion of our studies and our curiosities. Nor would it be a question of making a simple choice in its favor. For the historian, accepting the longue durée entails a readiness to change his style, his attitudes, a whole reversal in his thinking, a whole new way of conceiving of social affairs. It means becoming used to a slower tempo, which sometimes almost borders on the motionless. At that stage, though not at any other . . . it is proper to free

oneself from the demanding time scheme of history, to get out of it and return later with a fresh view, burdened with other anxieties and other questions. In any case, it is in relation to these expanses of slow-moving history that the whole of history is to be rethought, as if on the basis of an infrastructure. All the stages, all the thousands of stages, all the thousand explosions of historical time can be understood on the basis of these depths, this semi-stillness. Everything gravitates around it. (*On History*, p. 33)

Convinced as he was that the longue durée was essential to the understanding of life, Braudel nevertheless retained a notion of time similar to Marc Bloch's, a notion of a single, universal, and continuous time that comprehended all events alike and was the touchstone of all true historical study. "For the historian everything begins and ends with time, a mathematical, godlike time, a notion easily mocked, time external to men, 'exogenous,' as economists would say, pushing men, forcing them, and painting their own individual times the same color: it is, indeed, the imperious time of the world" (*On History*, p. 48). This lingering suspicion that the time of the longue durée was not true historical time worked its influence on Braudel's own historical writing. In his last major work, a three-volume study of civilization and capitalism in the early modern period, he considered only the third volume, with its chronological focus, to be true history. "The three volumes that make up this book are entitled: *The Structures of Everyday Life: The Limits of the Possible*; *The Wheels of Commerce* and *The Perspective of the World*. The third is a chronological study of the forms and successive preponderant tendencies of the international economy. In a word, it is a *history*. The first two volumes are much less straightforward, and come under the heading of thematic research." [12]

But the historical time that Braudel considered so fundamental was only one particular historical time: Newtonian time. Before the seventeenth century, historians did not have such a concept of time and did not think one time could comprehend all historical events. In the chapters that follow I will explore the ramifications of pre-Newtonian time in the practice of history in the West. Examples of stories told in relative time abound in non-Western literature and historical writing. Studies of the American Indians, beginning with Whorf's on the Hopi, have identified a sense of time much closer to that which underlies the Einsteinian universe than the one Westerners currently use in everyday life. [13] Chinese historians do not assume the existence of an overarching time series independent of the events. They prefer to date from specific dynasties.

These non-Western narratives by their very nature are hard to incorporate into our own experience. The sense of absolute time seems a distinctive Western contribution, coloring our view of the world and shaping our sense of self and society. Absolute time is undeniably a Western contrivance, but most Western history is not recorded in absolute time. The historians discussed here are not remote from our world; they described many of the events that form the basis of our identity. The clash between Europe and Asia at the dawn of Western history, the achievements of Athens in the Golden Age of Greece, the rise and fall of Rome, the coming of Christianity, the growth of the territorial state, and the birth of European nations, all of these were conceived in relative time. All bear the marks of a temporal orientation different from our own.

In telling these stories the historians used dating systems appropriate to the event being recorded. New dating systems were devised in order to express new historical identities that demanded the synchronization of events which had had no temporal relationship in earlier chronologies. For the purposes of this book it does not matter whether these new identities actually occurred in datable time—like the rise of Rome—or whether it was only the perception that occurred in time—like Petrarch's discovery of the personality of the ancients that gave rise to a new sense of the relation between the ancient and modern worlds. In either case the result was the same. Since the new systems were adopted to serve a specific new reality, the times they dated did not always apply to other sorts of events, which often continued to be measured by older, established chronologies. In order to explain the temporal relationship among events historians had first to make judgments about their meaning, importance, and direction. The dating systems the historians used derived their value from the events they were measuring; the dates did not have the universal significance that we accord to ours. They did not express absolute temporal location.

In selecting writers for treatment I have followed a variety of principles. Those who wrote self-consciously in the historical genre and who have been generally recognized as historians seemed more appropriate for study than those whose reputations were more poetic or metaphysical. Since my argument stresses the affinities between history and literature, it seemed unwise to confuse that issue by including writers whom subsequent generations have not regarded as historians. Thus, though Virgil has certainly contributed much to the sense of time that dominated the Oecumene and the Christian Middle Ages, I have preferred to discuss Livy and Pompeius Trogus as writers of the Augustan period. Similarly I have chosen Herodotus over Homer, Augustine over the Prophets, and Guicciardini over

Shakespeare. Similarly, except for Chapter 2, where metaphysicians played a key role in the processes under discussion, philosophers have been avoided, unless they made direct contributions to chronological technique.

Even among the historians selectivity was necessary, and several considerations have determined the choice of historians for this study. Some, like Herodotus, Thucydides, or Tacitus, are obvious historical classics. They have defined the standards of good historical writing for centuries of readers and have served as models to their successors. Others, like Augustine or Machiavelli, are of less significance for their historical writing than for their ideas about time and history. Augustine, for instance, formulated conceptions of progressive time that have distinguished Western thought. Still others, like Orosius or Pompeius Trogus, lacked the historical sophistication of Tacitus or Augustine but summarized events for subsequent writers so effectively that their writings were widely read during significant periods and should not be ignored. Thus some are chosen more for their historical and conceptual sophistication; some for the breadth of their influence; still others for both.

The scholarly literature in this field is as contradictory as it is vast. One can make few generalizations about any of these historians or about the field in general without taking sides in scholarly disputes. I have allowed these disputes to intrude as little as possible, discussing them only where I thought it would illumine a particular problem. Careful readers will notice the influence of some scholars more than others. Hayden White's *Metahistory*, with its analysis of the relation of historical consciousness, narrative, and knowledge in the nineteenth century, has helped me formulate the issue for an earlier tradition, and Erich Auerbach's *Mimesis* has provided a model for the analysis of historical narrative. G. W. Trompf's remarkable synthesis of Western ideas of recurrence has been most helpful in assessing that important aspect of historical time. Certainly Virginia Hunter and Peter Stahl have influenced my judgments on the Greek historians; T. J. Luce and F. Walbank on the Hellenistic period; R. W. Southern for the Middle Ages; and Mark Phillips and William Bouwsma on the Renaissance. In general, historians who look at the internal structure of a work and assess its values and preconceptions are more useful to my purposes than those who simply consider its accuracy and the extent to which it can be used by historians working according to the canons of modern historical scholarship.

Pre-Newtonian chronologies and dating systems have not gone unremarked in modern scholarship. Studies abound, but almost without exception they are interested mainly in how to translate the dates in these works into the modern chronology. Some of this work is extremely useful, setting out the techniques of the system with great clarity and explaining its peculiarities in relation to the B.C./A.D. system. Bickerman's study of ancient chronology is clear and comprehensive, as is Grumel's on world eras of the medieval period; more particular studies such as Philip Deane's work on Thucydides' Pentekontaetia, Mosshammer's on Eusebius, Jones' on Bede, or Grafton's on Scaliger, are especially helpful. Though my own work could hardly have proceeded without these scholars, the questions posed here are fundamentally different from those answered in other studies of chronology. I am not directly interested in the extent to which the dates in Herodotus or in Bede can be expressed in absolute dates of the Newtonian system. Instead I want to know the relation between the dating systems chosen, the assumptions the authors made about time, and the type of event they described.

In a work of this scope complete consistency of citation remains an elusive goal. Though all works have been consulted in the original language in the editions cited, I have tried to use readily available translations wherever possible. In the case of classical historians, the best modern translations usually relegate the dates to footnotes or eliminate them entirely, making those translations inappropriate to my purposes. In those cases I have used translations from the Loeb editions. Where modern translations are unavailable or inadequate, I have supplied my own. In general I have cited common primary sources by chapter and book rather than page number, so that any available edition or translation may be consulted.

The focus of this study is definitely on the broad historical tradition that stretches from Herodotus to Newton. After a first chapter which describes the historical genesis and decline of absolute time, five chapters will present the major dating systems in use during this period, discussing the historical background which produced them and the manner in which they were combined with earlier systems. A seventh chapter explores the differing standards of historical truth that obtain in a tradition of historiography where things cannot be unequivocally located on a single time line. Despite this focus, it remains my hope that acquaintance with the historians who wrote before Newton will take on added meaning in context with changes in time and space that are occurring in our own period. The epilogue will contain some reflections on the affinities between the approach to human experience recorded here and the tentative approaches toward a new description of such experience in this century.

2

The Rise and Fall of
Absolute Time

I do not define time, space, place, and motion as being known to all.
Only I must observe that the common people conceive those quantities under no other notions but from the relation they bear to sensible objects. And thence arise certain prejudices, for the removing of
which it will be convenient to distinguish them into absolute and
relative, true and apparent, mathematical and common. Absolute,
true, and mathematical time, of itself, and from its own nature flows
equably without regard to anything external, and by another name is
called duration. Relative, apparent, and common time, is some sensible and external (whether accurate or unequable) measure of duration by means of motion which is commonly used instead of true
time; such as an hour, a day, a month, a year.

Isaac Newton, *Mathematical Principles
of Natural Philosophy*, bk. 1, def. 8, scholium

Notions of time and space are so basic to our experience of life that they
tend to take on the qualities of common sense, to become such a natural
aspect of things as to lie beyond question, immune from the ordinary
changes that other ideas and values must suffer. Many are surprised when
they first encounter this passage from Newton's *Principles* and discover that
the time to which his laws of motion apply is not the astronomical and
relative time of the sun and the moon but a pure, abstract, and absolute
time which extends infinitely over all possible experience, a time that is
universal, continuous, and completely without dependence on any natural

regularity. Though Newton's concept of absolute time may seem unnatural when first encountered, it has in fact become embedded in many of the notions we apply automatically to the physical world. These notions constitute what we mean by common sense even today, nearly a century after experiments first showed the concept behind them to be an inadequate construct for experience.

Our ordinary notions of time and space spring from a complex historical process. To describe this process, to tell the story of the genesis, maturity, and decline of absolute time and space, requires the elucidation of unusually recondite metaphysical issues. The task lends to this chapter a degree of abstraction considerably greater than that which will characterize the book as a whole. This level of abstraction arises from the comprehensiveness and cogency of the metaphysical system that emerged during the eighteenth and nineteenth centuries, a system founded on the concepts of absolute time and space.

The ensuing discussion will seek to clarify three major points in hopes of producing a sympathetic appreciation of those constructs of time and space that supported historical writing before Newton. First, absolute time and space were crucial to the certainty implicit in scientific methodology, but to obtain this certainty and create a metaphysical system that applied to all possible experience thinkers had to solve difficult epistemological problems. The solution to this question of how we know the truth was so compelling that it became difficult to see phenomena that did not fit into the system. Only in the late nineteenth century did physicists find empirical data that could not be adequately explained with the constructs of absolute time and space.

Second, the epistemological problems associated with absolute time and space revolved around the issue of individual identity and the place of the observer in a world of changing events. To preserve certainty the observer had to be conceived as fixed, indivisible, and absolute. This construct became necessary to the system but defied certain aspects of subjective experience. Here again the cogency of the system prevented a critique of this point until the late nineteenth century when Freud—as well as many artists and writers—postulated deep ambivalence and paradox within the individual.

Third, the historical profession has a particularly close and intimate relation to the development of this metaphysical system. Modern historical methodology depends largely on the assumption that general historical events and processes are made up of smaller events that can be directly documented from primary sources. Primary-source criticism became the foundation stone of historical methodology contemporaneously with the

Kantian solution to the metaphysical problems inherent in absolute time and space. History's claim to objectivity and a place among the social sciences depends in large measure on perceiving concrete historical events in the way nineteenth-century physicists perceived atoms—as the constituent elements of nature and of the past.

This third point needs to be carefully understood. The dependence of historical methodology on absolute time and space should not blind us to the richness and flexibility of modern historical writing. No great historian has limited the historical inquiry strictly to those simple primary events that can be definitely located in time and space. Motives, large cultural and political identities, secular trends all play an important role in modern historical narratives during this period. But they do not fit easily with the historical methodology. To incorporate them into their picture of the past, historians often had recourse to techniques of narrative and discourse drawn from an earlier historical tradition, one that did not accept the assumptions on which the historians' claim to accuracy was based.

Leopold von Ranke, who as much as any other established this method as the dominant mode of historical inquiry, demonstrated in his theoretical writings the importance of objectivity and strict accuracy that he prescribed for history. By his death in 1886 Ranke was widely acknowledged both in Europe and America as the father of modern historical writing. His first work, the *History of the Latin and Germanic Nations from 1494 to 1514*, proposed a new goal for historians. In the preface to that work, published in 1824, he said, "To history has been assigned the office of judging the past, of instructing the present for the benefit of future ages. To such high offices this work does not aspire: It only wants to show what actually happened [wie es eigentlich gewesen]."[1] Present-day readers can hardly appreciate the novelty of such a claim in the early nineteenth century. Not only had most of Ranke's immediate predecessors regarded history as primarily moral and judgmental, but he was speaking against a tradition going back to the earliest historians in the West. What could have given Ranke the confidence to oppose such a venerable tradition?

Ranke's confidence was first of all based on his method of research. He went on in the same passage to describe the sources on which he built his description of the past as it actually was.

> But whence the sources for such a new investigation? The basis of the present work, the sources of its material, are memoirs, diaries, letters, diplomatic reports, and original narratives of eyewitnesses; other writings were used only if they were immediately derived from the above mentioned or seemed to equal them because of

some original information. These sources will be identified on every page; a second volume, to be published concurrently, will present the method of investigation and the critical conclusions. . . . The strict presentation of the facts, contingent and unattractive though they may be, is undoubtedly the supreme law. (in Stern, p. 57)

Ranke expressed here his determination to establish the primary facts by exploring the sources of information most immediate to the events themselves. He felt that only after he had established these primary facts could he hope to explore larger issues of cause and pattern in history. Ranke did go on to consider these large issues, and in his later life focused his attention more and more fully on the unities which might lead to an understanding of universal history. But he always considered the primary fact to come first. The historian must not only establish when and where an event happened; he must show his reader how he came to this judgment. In that way the reader can achieve the same objective certainty that the historian himself had about the fundamental units of historical analysis.

In Ranke's eyes, then, the moral function of history gave way to a new standard of accuracy. This standard of accuracy in turn assumed that the events which the historian described were primary facts which could in theory be located unequivocally without reference to any other events. Events could not depend on other events for their dates, for then their meaning would come from a relationship which already existed before the historian dated them. The interrelatedness of events had to grow out of their particular existence; to reverse the process would run counter to Ranke's insistence on establishing the primary fact before identifying larger patterns. Modern readers assume that we can fix a date for the battle of Waterloo before determining whether it occurred after the French Revolution. The assumption appears to be simple common sense. Yet what is common sense to us would be strange to earlier historians, for the idea that events can be simply located in time is itself the product of a particular historical process.

In telling the story of this process, I have adopted a chronological organization that reflects the self-consciousness with which the thinkers involved built on the work of their predecessors. The organization is in part artificial, since the focus is less on linear progress than on the three points mentioned earlier—the centrality of absolute time to scientific method, the importance of the individual observer absolutely separated from the events being perceived, and the dependence of primary source criticism on the idea that the things that constitute the historian's subject matter can be simply and unequivocally located in time. If the reader carries away from this chapter an understanding of those points, its purpose has been served.

THE BIRTH OF ABSOLUTE TIME

The idea of simple location can be traced back to the work of René Descartes (1596–1650). Convinced that the traditional paths to learning would never lead to certainty, he set out on his own inquiry into the truth by putting aside all preconceptions and conventional modes of thought. In their place he sought to establish a system which he could prove by the clearest and most self-evident reasoning alone. Descartes felt his most reliable guide in establishing unequivocal truths would be geometry, whose proofs rested on pure reason and deduction from self-evident postulates. By 1637 he felt confident enough of the results of his inquiry to publish a description of his method. In this *Discourse on Method*, he acknowledged his debt to the geometricians. "These long chains of reasoning, each of them simple and easy, that geometricians commonly use to attain their most difficult demonstrations, had given me an occasion for imagining that all the things that can fall within human knowledge follow one another in the same way and that, provided only that one abstain from accepting anything as true that is not true, and that one always maintains the order to be followed in deducing the one from the other, there is nothing so far distant that one cannot finally reach nor so hidden that one cannot discover."[2]

Descartes wanted to use this geometrical method to analyze all possible aspects of experience, but he felt that such extended use would produce a system too cumbersome to remember or to picture. To solve this problem he invented analytical geometry. Since the continuum of numbers is infinite, he reasoned that he could use numbers to describe every geometrical object. (As in so many cases, Descartes, though the most famous name associated with the discovery of analytical geometry, was not the first. Fermat anticipated him by eight years.) He could express plane figures with two coordinates; solid figures, with three. An equation based on three variables could locate any object in a three-dimensional field. Simple geometrical forms enabled Descartes to understand basic relationships, and he could build complex groups out of these simple relationships by using algebraic equations.

> Now, having noticed that, in order to know these proportions, I occasionally needed to consider each of them individually, and

sometimes only to remember them, or to gather up several of them together, I believed that, to consider them better in particular, I ought to suppose them as relations between lines, since I found nothing more simple, nothing that I could more distinctly represent to my imagination and my senses; but to remember them or grasp them all together, I would have had to explicate them by means of certain symbols, the shortest ones possible; and by this means I would borrow all of the better aspects of geometrical analysis and algebra, and I would correct all the defects of the one by means of the other. (*Discourse*, p. 11; *Oeuvres*, 6:20)

Analytic geometry revealed to Descartes an exact relationship between number and space. Using the Cartesian coordinates, he could locate any object, no matter how large or small, with reference to an arbitrary point. No other object needed to be used in this process. Perhaps the analytic also implied an exact relationship between number and time to match that between number and space. Could it use numbers to describe events in time with equal precision and independence from other events?

Before Descartes could apply number to time he had to solve an important problem. Duration was not one of those things he could easily form a clear and distinct idea of. He could not even quantify the force which produced his own endurance. As he expressed the problem in his *Meditations*, "For the whole duration of my life can be divided into an infinite number of parts, no one of which is in any way dependent upon the others; and so it does not follow from the fact that I have existed a short while before that I should exist now, unless at this very moment some cause produces and creates me, as it were, anew or, more properly, conserves me."[3] Continuity in time, then, could not be explained mathematically. The force which caused it transcended our experience and the possibility of quantification. Duration could only be understood by referring to the creative force of God. This creative force did not undermine the predictability of Descartes' system, since God established the order of the universe by necessity and not by whim. Nevertheless, duration was completely dependent on it at every instant and could not be conceptually abstracted and quantified.

To overcome this obstacle and make possible the union of number and time, Descartes drew an important distinction between time and duration. Duration he regarded as simply that aspect of a thing we refer to when we speak of its continuing to exist. Time is limited to measuring motion; it is distinct from duration. Since time is only a way of thinking about duration, it adds nothing to our understanding of the continued existence of objects. The fact that we can find units to express how long something endures does not mean that we can use these units to understand the phenomenon of endurance itself. Furthermore, he argued, we can think duration only by means of a comparison. The units of time we use are simply common measures of different durations.[4]

Descartes' distinction between time and duration made it possible to think of absolute units of time, as long as it was clearly realized that these units added nothing to an understanding of duration. For him the spatial coordinates afforded by analytic geometry were so important that he did not regard the problem of discontinuity in time as of major significance. He attempted no chronological studies on a scale with his analysis of space. Isaac Newton (1642–1727) was far more concerned with the problem of time and constructed a system in which time was a clear, quantifiable, and independent factor.

Newton based his system on a concept of absolute space and time in which matter moves according to regular laws. He saw that he must set aside common notions of time and space and conceive duration itself in such a way that it could be directly quantified.[5]

> I do not define time, space, place, and motion as being known to all. Only I must observe that the common people conceive those quantities under no other notions but from the relation they bear to sensible objects. And thence arise certain prejudices, for the removing of which it will be convenient to distinguish them into absolute and relative, true and apparent, mathematical and common. Absolute, true, and mathematical time, of itself, and from its own nature flows equably without regard to anything external, and by another name is called duration. Relative, apparent, and common time, is some sensible and external (whether accurate or unequable) measure of duration by means of motion which is commonly used instead of true time; such as an hour, a day, a month, a year. . . . Absolute space, in its own nature, without regard for anything external remains always similar and immovable.

Absolute time was as essential to Newton's system as absolute space, for without duration he could not establish the laws of physics. We cannot perceive absolute space, since it is empty; we can only perceive the absolute motion of objects in space. This absolute motion is continuous, and bodies in absolute motion produce a centrifugal force which can be measured without reference to any other body. Thus Newton used the concept of absolute time as the metaphysical foundation by which absolute space was made accessible to empirical knowledge.

But Newton's interest in time went beyond the problem of physical motion. He studied deeply in the chronology of human events and felt that his mathematical system could reduce these to the same quantifiable order that he found in the physical universe. To this end he wrote a series of chronological works, including the *Abstract of Chronology* (1726) and the *Chronology of Ancient Kingdoms Amended*, published in 1728, the year after his death. In these works Newton was less interested in the human problems posed by the study of the past than in dating past events precisely. "Newton's historical studies were always aimed at reducing events, persons, and objects which were by tradition surrounded with a vague mystical aura to matters of fact, preferably to mathematical terms. He had precious little interest in historical character or motivation. To know a quantity and an exact date was one of the ultimate goals of his realistic history. In the end his passion for factual detail shriveled the past to a chronological table and a list of place names."[6]

No large historical insights or new historical trends emerged from Newton's work. He relied on the traditional historiographical framework of his day for the patterns he saw in history. But his work has a significance beyond the realm of historical theory, for he presented in it an implied temporal continuum stretching infinitely forwards and backwards in terms of which any event could be precisely and simply located. Political realities, no less than physical ones, could be charted on this continuum. No event was too ancient, no duration was too long to evade its potential grasp. "In Newton's new system of chronology the world physical and the world political established a vital point of contact" (Manuel, p. 164). On this point of contact Ranke's historical method was founded.

THE ROLE OF THE ABSOLUTE SUBJECT

Newton's chronological work did not immediately produce a historical method based on the absolute time line. Nearly a century passed before Ranke taught his first research seminar; the *History of the Latin and Germanic Peoples* appeared almost two centuries after Descartes' *Discourse on Method*. Why did it take so long for the concept of absolute time to stimulate an appropriate historical method? The answer to this question lies less in the concept itself than in another set of ideas on which the concept was based. Thinkers during the eighteenth century encountered serious obstacles as they sought to create a single philosophical system which could use the absolute time line to describe actual human experience.

The fundamental problem was epistemological. In order to explain how absolute space and time produced certainty Descartes was led into a dualism between the objective, knowable world and the thinking, knowing subject. Descartes insisted that these were two separate substances with entirely different properties. The knowable world was extended, material, and moving; the thinking subject was spiritual, immaterial, and stationary. This complete dualism was a precondition for objective knowledge. Descartes' division between thought and extension in turn raised an important question of how the two elements of this dualism were related, a question which dominated philosophy in the eighteenth century and which had to be answered in a reasonable way before the concept of absolute time could carry the conviction necessary for building a historical method on it.

How does the time line become enmeshed in such metaphysical issues? There is in fact a clear progression of thought from the idea of absolute time to the problem of epistemological certainty. In the last section we saw that absolute time was implied in the notion of absolute space, since only in motion could objects be measured in absolute space. Absolute space in turn was an essential premise of Descartes' analytical geometry, which he felt would lead him to certain truth. But analytical geometry alone did not produce the certainty Descartes required. As much as he admired the geometricians, he felt they erred in not questioning the assumptions on which they worked. He wanted to find some philosophical assumption that would give objective certainty to his postulates about straight lines, angles, and geometrical figures.

Standing in the way of this certainty was the pluralism of seventeenth-century thought. The science of Descartes' day still accepted the Aristotelian notion that each field of inquiry had a science appropriate to it with its own separate methods and assumptions. Descartes was convinced that he could reduce all of these to a single verifiable premise by looking in the field of philosophy, to which all of the various sciences referred their basic ideas (*Discourse*, p. 18; *Oeuvres*, 6:21–22).

Descartes embarked on a search for this basic premise by a process of systematic doubt. He called into question not only all scientific ideas but all evidence of his own senses and all products of his own reason. This systematic doubt produced his most famous contribution to Western thought, the "cogito ergo sum." He found that his very capacity to doubt made it impossible to doubt his own existence, since he could not doubt if he did not exist. What Descartes had proved was not his whole existence but only his existence as a thinking being, apart from any material traits. "Thus this 'I,' that is, the soul through which I am what I am, is entirely distinct from the body, and is even easier to know than the body, and even if there were no body, the soul would not cease to be all that it is" (*Discourse*, p. 18; *Oeuvres*, 6:33).

Descartes felt he could prove the existence of the material world from the self-evident fact of his own spiritual existence. The fact that he could doubt made him imperfect, since to know was clearly more perfect than to doubt. Since imperfection was a negative trait, it depended on the existence of a perfect being. The existence of a perfect being in turn demonstrated that of the outside world, since a perfect being would have no reason to deceive us about such an important issue as the existence of matter. Descartes' conceptions of the material world could not spring from nothing. Specific observations of matter might be in error, but the ideas of matter he possessed must have been put into his mind by a being more perfect than he. The idea of such a being was as clear and distinct as his own existence and could only have been placed in him "by a nature truly more perfect than I was, and even that it had within itself all the perfections of which I could have any idea, that is, to put my case in a single word, that this nature was God" (*Discourse*, pp. 18–19; *Oeuvres*, 6:34).

This chain of reasoning from clear and self-evident truths led Descartes to the necessary existence of himself and the world around him. He existed as a thinking subject capable of reasoning to certain truth; he also existed as a body, part of the extended, objective world. But where does time fit into this picture? He was convinced that it was part of the extended world rather than part of his thinking self. It was not among those aspects of matter which were directly perceptible, aspects like taste, color, or sound, which he called secondary qualities. Duration was a primary quality; it could not be seen but was nevertheless an essential trait of all matter. Just as matter had magnitude, figure, and number, so it had duration and motion as inherent characteristics. These primary qualities were part of the idea of extension just as the idea of a triangle included the number of degrees in its angles (*Principes*, 4.198–199; *Oeuvres*, 9:316–18).

Since the senses could not grasp duration and motion directly, an event could be measured in time not so much because it existed as a sense object as because the idea of extension and duration existed in the mind of the knowing subject. These ideas were innate, neither traceable to nor verifiable by experience. Descartes did perform experiments to test some of his hypotheses, but he regarded such references to the empirical world as merely didactic, intended to convince people who could not understand the mathematics on which his ideas actually depended. When he discussed the circulation of the blood, for instance, he told his readers to consult Harvey's experiments, but only if they could not understand the force of his mathematical arguments for circulation (*Discourse*, pp. 26–27; *Oeuvres*, 6:50).

Time and space, then, exist in extended objects, but the temporal and spatial characteristics of these objects can be measured and analyzed into certain truths only by a thinking subject whose process of thought is not itself extended. Descartes' attempt to explain the connection between these two parts of being led him down many strange pathways. He tried to argue that the pineal gland was the organ where mind and body communicated with one another. He suggested whirling vortices as the source of motion in the universe. Even his distinction between time and duration is in part an attempt to bridge the gap between an extended world in which duration is real if not completely comprehensible and a thinking subject whose units of temporal measurement are absolutely certain.

Descartes' theories of the pineal gland and the vortices were little more convincing to his contemporaries than they are to us. Even those who shared his rationalism found his system too dualistic. Spinoza attacked his notion of two substances and constructed a system based on a single substance of which extension and thought were two attributes. Newton, who considered the force of gravitation itself to be occult and immaterial, did not feel that the relation between subject and object needed complete clarification. His concepts of space and time could stand alone as a useful basis for a predictive science.

But the problem remained. As impressive as Newton's laws of gravitation were, they still assumed a thinking subject whose conceptual picture of time and space made knowable the regularities which actually existed in nonthinking objects. As thinkers sought to apply these concepts to wider and wider fields of human experience, the lack of predictability in these fields would sooner or later call into question the very notion that thought could understand experience by using its own concepts.

Absolute time does not necessarily accompany the rationalist outlook. One of the most important rationalist thinkers of the period argued for a concept of relative time. Gottfried Wilhelm Leibniz (1646–1716) opposed both Descartes' dualism and Spinoza's monism with a pluralism based on monads. According to Leibniz, each of these tiny bits of mind reflected the order of the whole universe but did not interact with any other monad. It was Leibniz' very rationalism that led him to reject the notion of absolute time. He felt that for the rationalist view of the universe to be valid there must be a reason for all real aspects of experience. He showed that the doctrine of sufficient reason made absolute time impossible. If time were really part of things, then there must be a reason why the world was created exactly when it was and not a few years earlier. Since the total succession of events would be the same regardless of when the first event occurred, Leibniz argued that time was relative, expressing the succession of coexistence of real things, and that it did not have an existence apart from these things.

This relational concept of time seems to prefigure Einstein's time, but Leibniz had little direct influence in that particular field. His theory of monads was in general too farfetched and invoked too many notions outside the bounds of common sense. The idea of his that received the widest popular dissemination was the proposition that this was the best of all possible worlds, a philosophy viciously satirized in Voltaire's *Candide*. Leibniz exerted his most serious influence in physics, where he provided the basis for eliminating divine intervention from physical processes. Since each monad reflected the entire universe, God was not necessary to sustain the regularities. Various parts of the universe exhibited the same laws of motion for the same reason that two clocks keep the same time.

Far more challenging to the rationalist notion of absolute space and time was the empiricist critique. The empiricists questioned the very premise that the structure of the mind was a guide to the sensible world. Philosophers in the tradition, most notably Locke, Berkeley, and Hume, sought the source of human knowledge in the raw data of experience. They argued that the ideas present to the mind were themselves the product of experience. Drawing on an intellectual tradition going back to the nominalists of the twelfth century, they maintained that sense data alone could assure us of the existence of things. Our abstractions only help us understand; they do not show us that things actually exist.

Since time was an abstraction, its status within the world itself became less secure. All of those qualities of matter which Descartes considered primary and invisible to the senses could be traced indirectly to sense data. John Locke (1646–1704) divided all ideas into simple ones, where the mind receives data directly from the senses, and complex ones, where the mind combines various data into a concept. Our idea of the extension of a particular object is a simple idea produced by the sense of sight, but the idea of extension in general is a complex one. It arises when we combine a series of simple ideas of particular extended objects. In doing this we have done no more than repeat a number of simple ideas and give our repetitions a name, a name which refers to nothing actually existing outside our minds.[7]

Time is even further removed from actual events and more completely the result of exclusively mental processes. Duration arises from our observation of a succession of ideas in our minds and is simply the term we assign to the distance between the parts of that succession. The units of duration, like days or years, come from regularities we have experienced, while our notion of time itself is nothing more than the complex idea of these units abstracted from any particular experiential regularities. Time is thus a construct of the mind; it is not found directly in our experience nor can it be proved to be a part of the outside world (Locke, 2.14.31).

The attraction of Locke's ideas lay in their appeal to common sense. The rationalist attempt to construct a universe entirely out of abstract, analytical concepts had led to a system too remote from the everyday world. Distrust of such abstractions had been a long standing part of British intellectual life. A century before Locke, Richard Hooker had referred to excessive generalization as a "cloudy mist cast in the eye of common sense." This appeal to common sense alone justified abandoning the logical necessity which the rationalists claimed for their arguments. Geometry may produce apodictic truth, but the empiricists insisted that life was not a geometrical form.

The appeal to common sense was, however, a two-edged sword. If the mind was, as Locke insisted, a blank sheet at birth, all of whose contents could be traced to specific experiences, then all thoughts could be explained by reference to specific empirical data. In trying to do so Locke found himself casting his own cloudy mists in the eye of common sense. For Locke could not trace the notion of substance to any particular sensation. We perceive all of the secondary qualities of a thing through our senses, the primary qualities through a combination of senses. That is clear enough, but what sense tells us that the collection of sense data emanates from a thing and not just a disorganized field of unconnected sensations? This is certainly not a question to be easily ignored, since the idea that we are surrounded by things is one of the most basic elements of our common sense. Despite the importance of substance Locke could not find an experiential source for it. He called it a supposition which we are forced to make, a supposition of something, we know not what, in which our simple ideas of sensations subsist, since they cannot subsist in themselves (Locke, 2.23.1–2).

Locke's supposition of an occult material substratum undercut the empirical basis of his system and called into question his assertion that time is a mental process constructed out of sensations which have no intrinsic temporal relation. If we have to suppose something underlying our sensations which cannot be seen, how can we know whether this something has temporal duration? Going further, if such an important idea as substance is not empirically grounded, how can the empiricists defend their central proposition that all contents of the mind arise out of external sensations?

Attempts to produce an empirical philosophy not dependent on an occult material substratum tended to diverge even more completely from the dictates of common sense. George Berkeley (1685–1753) argued that it was only Locke's assumption of the existence of matter that forced him to suppose an occult material substrate. In fact nothing in our experience requires that assumption. We are conscious of sensations only as particular ideas. Only by intellectual carelessness do we attribute a material basis to them. Berkeley reduced experience to three components: finite ideas, finite spirits that perceived the ideas, and God, an infinite spirit whose ability to think all the ideas kept them in existence when no particular finite spirit was thinking about them. His system is annoyingly logical, hard to attack, but defies common sense at every turn.

By the middle of the eighteenth century it was hard to maintain either that time was an absolute reality or that it was only a mental construct. Both assertions seemed logically and practically flawed. No thinker pointed up the difficulty more sharply than David Hume (1711–76), the most thoroughgoing empiricist Europe had seen since William of Occam. Hume argued that we cannot assess the truth of our propositions until we understand what they mean. Propositions do not mean anything at all unless they can be verified. Empirical propositions concern matters of fact and can be verified by asking what simple sensation or impression produced them; mathematical propositions concern relations between ideas and can be verified by the law of noncontradiction. Propositions like "God exists," "substance is a necessary postulate of our experience," or "matter exists," are neither true nor false; they are simply meaningless.

Hume did not limit his critique to these metaphysical propositions. He went on to call into question a major premise of seventeenth- and eighteenth-century thought, the notion that every effect has a cause. This idea is the basis of scientific knowledge, for if there are indeed uncaused events then the predictive hypotheses of science lose their necessary validity. Hume argued that the idea of cause does not come from our experience. No matter how carefully we analyze our experience we see only one event happening after another. The two events are conjoined but not connected. Nor is the proposition that every effect has a cause an analytical one, for its denial does not involve a contradiction. We cannot disprove the statement "there are uncaused events" in the same way we disprove the statement "a triangle has four sides."

This critique does not so much solve the problem of time as make it irrelevant. Hume reduced time to an impression that sensations succeed one another. Nothing in that impression allows us to generalize from these sensations to make any necessarily true statements. We cannot even make necessarily true statements about that particular succession, to say nothing about duration in general. The idea that events could be simply located in time had lost both its logical validity and its empirical basis. Absolute time had become a useless and meaningless metaphysical speculation.

As a final blow to the rationalist notion of time Hume also attacked the idea of the absolute subject. For Descartes the absolute separation of thinking subject from material object made possible the perception of things in absolute time. Hume argued that the self can no more be traced to a simple perception than can cause and effect. "For from what impression could this idea be derived? . . . For my part, when I enter most intimately into what I call myself, I always stumble on some particular perception or other, of heat or cold, light or shade, love or hatred, pain or pleasure. I never can catch myself at any time without a perception, and never can observe anything but the perception." Hume deprived the mind of its unity. No longer independent from the world of sense objects, it became a series of sensations. He likened it to a theater that does not exist except for the plays that are enacted there. "There is properly no simplicity in it at one time, nor identity in different; whatever natural propensity we have to imagine that simplicity and identity."[8]

Hume showed in the most dramatic terms the inherent weaknesses in Cartesian dualism. If the world is divided into the thinking subject and the material object, neither rational nor empirical evidence can suffice to join these two opposites and produce a single system depending on a single postulate. He stated the basis of his attack eloquently at the end of his *Enquiry Concerning Human Understanding.*[9]

> When we run over libraries, persuaded of these principles, what havoc must we make? If we take in our hand any volume; of divinity or school metaphysics, for instance; let us ask, "Does it contain any abstract reasoning concerning quantity or number?" No. "Does it contain any experimental reasoning concerning matter of fact and existence?" No. Commit it then to the flames: For it can contain nothing but sophistry and illusion.

At the beginning of this section the question arose as to why it took two hundred years to create and implement a historical method based on the possibility of accurately locating an event in absolute time. In the empiricist critique an answer has emerged. By the middle of the eighteenth century serious problems had developed in the attempt to create a co-herent and complete philosophical system based on the premise of abso-lute time. Most importantly, absolute time depended on a strict division between subject and object that could not easily be defended. Neither rea-son itself nor the sense data of everyday experience could adequately de-scribe the universe using the notion of absolute time and the absolute thinking subject. Since both of these were essential parts of any scientific system which presumed to locate experience unequivocally, historians could hardly use them in the concrete task of describing actual historical events.

THE KANTIAN SOLUTION

To meet Hume's devastating critique, a thinker would have to reconcile the sharp division Descartes had made between subject and object. The reconciliation, however, should not sacrifice the objectivity and apodictic certainty which Descartes' system had afforded. The thinker most influen-tial in working this resolution is Immanuel Kant, whose philosophy of transcendental idealism brought subject and object together in a new way. He preserved the objectivity of absolute time while limiting its use to those realms of existence which the natural scientist and historian found most useful. Only by limiting its scope could he protect it from the ravages of metaphysical speculation that had proved so damaging in the eigh-teenth century.

Kant found a way out of Hume's challenge by showing that the British empiricist had not gone far enough in his critique of abstract reasoning. Hume's confidence in abstract reasoning lay in its analytical character. Since in analytical propositions the subject includes the predicate, error can be clearly identified through the law of non-contradiction. Kant at-tacked Hume on precisely this point, showing that the world of number relations, which typified for Hume the certainty afforded by analytic prop-ositions, was not in fact based on the law of noncontradiction. In the proposition "$7+5=12$," no matter how closely we analyze the subject, the number 12 is not found in it; some operation of the mind adding 7 and 5 together is necessary. Since the apodictic truth of the proposition is ob-vious, Kant argued, we must agree that some synthetic statements—state-ments where the predicate adds something to the subject—are true a pri-ori, independent of all experience.[10]

Kant had changed Hume's question fundamentally. We need no longer ask if synthetic a priori statements were possible; we must ask how they are

possible. To this new question Kant answered that synthetic a priori statements, like those in arithmetic and natural science, were possible only if
we make a crucial distinction between things as they are in themselves
(noumena) and things as we experience them (phenomena). Statements
acquire their necessary truth because they apply only to phenomena. Phenomena, that is to say experience, contain not only information from the
outside world but also elements that are purely mental, existing in the
mind apart from any experience. By making ourselves aware of those
things that exist in the mind before experience, we can discover how it is
that some synthetic judgments can be true a priori and not depend on
sense perception of the world outside the mind. Such synthetic judgments
may be true of the noumena or they may not; we cannot know since we
cannot experience things as they are in themselves (*Prolegomena*, pt. 1,
sec. 10, pp. 30–31; *Critique*, bk. 2, chap. 3, pp. 187–202).

Things exhibit an order not because of the way they are in themselves
but because our minds contain the notions of time and space and apply
these notions to experience as a necessary part of the act of perception.
Time and space are thus not derived from the outside world; before all experience they exist in our minds as a priori intuitions. Neither are they
part of the reasoning process. They are simply particulars that exist in the
mind, whose function is to order our experience. They constitute, as it
were, the glasses through which we perceive the world. Since we cannot
remove the glasses, we cannot know what the world around us really looks
like in itself; we can perceive it only in time and space (*Prolegomena*, pt. 1,
sec. 7–13, pp. 28–34; *Critique*, 1, pt. 1, pp. 19–49).

Kant, then, agreed with Hume that time did not exist in the external
world, but he did not think it was subjective. He argued in fact that this
very separation from the external world offered the only certain means of
making time objective. By making time an a priori intuition of the mind, a
pure form for the organization of experience, Kant turned it into a necessary part of all perception of experience, no longer dependent on the vagaries of the senses or the peculiarities of individual observers. "My doctrine of the ideality of space and of time, therefore, far from reducing the
whole sensible world to mere illusion, is the only means of securing the
application of one of the most important kinds of knowledge (that which
mathematics propounds a priori) to actual objects and of preventing its
being regarded as mere illusion" (*Prolegomena*, pt. 1, remark 3, p. 89).

With this distinction Kant substantially changed the meaning of objectivity. Objectivity no longer meant conformance with external things. Instead it referred to the universal validity of intuitions essentially subjective
in origin. He argued that space was clearly one of these intuitions, since

we could think of empty space but not of objects existing without space. Time was even more universal. Space was a necessary intuition only for phenomena of the external world. Time was an essential ingredient in all phenomena. Our feelings can change from fear to hope without moving in space, but we cannot perceive the change except in a temporal succession (*Critique*, 1, pt. 1, sec. 2, pp. 30–32). Thus Kant, abandoning the existence of time in the noumenal world, made it absolute and objective within the phenomenal one. It became a necessary precondition to experience, allowing reason to reach conclusions which were objective as long as we did not try to apply them outside the realm of possible experience. Statements like "space has three dimensions" or "time has only one dimension" were necessarily true even though they originated in subjective activity.

Kant's distinction between noumena and phenomena avoided the problems of Cartesian dualism. The process of understanding the material world goes on entirely within the mind. We are no longer faced with an irreconcilable split between thought and extension. At the same time the distinction avoided the obvious difficulties of empiricism. While Kant accepted the data of the senses as the source of our perceptions of the outside world, he insisted that knowledge is an active process of the mind imposing upon these data an order which makes them comprehensible. Whether things in themselves actually contain the order our understanding imposes on them is not a meaningful question, since we can only experience things as phenomena and will never know what order they may have in themselves.

At the same time Kant was far from denying the reality or importance of the noumenal world. Not only does the noumenal world present the data which our judgment makes into experience, but the very subjective activity which produces these judgments is noumenal. Where Descartes had seen a thinking subject in an extended body, Kant saw a single subject. It was on the one hand noumenal because it was active, but on the other hand it was phenomenal because its activity was observable in space and time. Kant made the subject the point of contact between the noumenal and phenomenal world, giving it both existential status within phenomena and logical status as a noumenon.

Kant's position on the noumenal subject is a complex one. Essentially he showed that the processes of reason gain the completeness they require only by identifying the ultimate premises from which they draw deductions. Descartes' identification of the thinking subject as the ultimate premise of all existential statements was correct, but Kant felt he attempted to draw conclusions from this premise which were inappropriate.

Since the subject contains the a priori intuitions of space and time in which all phenomena are found, we cannot apply concepts of space and time to it. We cannot ask if the subject is permanent, simple, or unified, for these are concepts which apply only to possible experience. Kant would thus agree with Hume in regarding a question like "Does the soul survive the death of the body?" as meaningless. Since we only experience the soul in life, death is the end of the soul as an object of possible experience, and time is an intuition applying only to possible experience (*Prolegomena*, pt. 3, sec. 45, pp. 80–81).

What then is the use of the transcendental idea of the ultimate subject as the ground of experience? According to Kant, the subject, or Ego, makes a strictly materialist point of view impossible. Since the noumenal subject is the condition of our consciousness, all the observable characteristics of the material world depend for their perception on a priori intuitions which are not themselves material. All objects exist in the three dimensions of space, but space itself has no dimensions and is not a spatial object. All phenomena exist in a single dimension of time, but time itself is without duration. The possibility of experience is dependent not on a material substratum but upon the unification of perceptions by a thinking subject. The noumenal subject is Kant's final critique of Cartesian dualism (*Critique*, Second Division, bk. 2, chap. 1, pp. 269–94).

Kant gave the absolute subject and the absolute temporal continuum a firm logical and metaphysical foundation. To be sure, the lack of such firm foundation had not prevented natural scientists from using absolute space and time. Before Kant, however, historical narrative proved resistant to these concepts. Observed patterns in natural phenomena are potentially verifiable by controlled experimentation. The historian can not easily obtain the repetition on which such a technique of verification depends. Since actual historical events cannot be repeated, the narrator must be even more convinced than the natural scientist that the assumptions which govern their temporal existence are valid. Before Kant such conviction eluded the grasp of critical thinkers. Kant resolved the metaphysical problems which stood in the way. In so doing he provided a rational system explaining both the objective validity of absolute time and the necessity for an absolute perceiving subject to organize events on the time line.

Kant's metaphysics led to both positivist and philosophical history. Historians like Ranke, who were primarily concerned with researching the past for its own sake, and philosophers like Hegel, who were more interested in commenting on the conceptual structure of history, based their historical work on Kant's premises about absolute time and the thinking subject. In order to understand better the place of these assumptions in modern historical writing, let us briefly consider these two approaches to history.

Kant did not at first apply his insights to history. More concerned with the natural sciences, he developed in his early work twelve categories of the understanding which he claimed would explain all scientific propositions about experience. Only later, after reading Herder's *Ideas for a Philosophy of History*, did he turn to consider the implications of his thought for that field. Even then he sought mainly to show how and why history conformed to natural laws. He argued that historical events must be subject to these laws to the extent that they were phenomena. Since the human acts themselves, conceived as noumenal, were the result of a thinking, moral subject, the laws underlying history should reveal some pattern of rational progress. Kant considered such a pattern difficult to perceive but potentially knowable. Kant, then, as might be expected, adopted a philosophical approach to history, more concerned with analyzing its pattern than discovering the details that illustrate the pattern.

In sacrificing the noumenal uniqueness of historical events to the demands of natural science Kant had created a philosophy of history with limited appeal to historians. Successors to Kant overcame this problem by redefining noumenal reality. Adopting Kant's insight into the active power of consciousness in organizing our experience, they insisted on the noumenal reality of consciousness as the ultimate subject, or Ego, which transcends objectification. Since the noumenal reality became intelligence-in-self rather than thing-in-self, these thinkers were called idealists, and their idealism reached the extreme of identifying the whole of reality with consciousness.

The reasoning behind this position can be seen in the works of the first major German idealist, Johann Gottlieb Fichte (1762–1814). In *The Basis of the Entire Theory of Science*, Fichte began with Kant's proposition that all judgments arise from the activity of the Ego or self. He went on to argue that this activity cannot be traced to a noumenal stimulus outside the consciousness. In trying to think about any object and to reduce that object to its ultimate reality, we are led to the activity of thought that produces it. The first proposition of all science is, then, the existence of the unlimited self. The objective world, or non-ego, is produced by consciousness only as the antithesis of self. It is the second logical postulate derived from the activity of consciousness, not a product of sense data. Our ideas of particular things arise only out of the confrontation of unlimited self and unlimited nonself. The existence of things is the third logical postulate of the activity of consciousness.

Georg Hegel (1770–1831) made Fichte's abstract and purely philosophical idealism into a comprehensive system which had direct relevance to the study of history. He accepted Fichte's notion that reality was a dialectical process of contradictory concepts and built a logical system out of triads in which two categories were opposites, to be synthesized in a third. On the most abstract and logical level he observed that Being itself, if we try to think it purely and without any limiting characteristics, becomes Nothingness, or not-being. The process by which Being turns into Nothingness is a synthesis of the two concepts in Becoming. "Their truth is therefore this movement, this immediate disappearance of the one into the other, in a word, Becoming."[11]

By incorporating Becoming into the fundamental triad of logic Hegel had made dialectical movement the ground of actuality. He gave a unique role to the study of history by distinguishing clearly between movement in space, which constituted the field of natural science, and movement in time, which was the province of history. "World history in general is the development of Spirit in Time, just as nature is the development of the Idea in Space."[12] Time is real and objective not only because it constitutes the framework within which the subject organizes possible experience; it is real because it is the process which exhibits the reality of the subject itself. Since the subject was the rational self, history displayed the development of reason. Hegel insisted that in history, as in the whole realm of being, what was real was reasonable and what was reasonable was real.

In making this statement he did not think he was forcing history into any preconceived mold or abandoning study of the particular events of the past. The historian still had to investigate the past in all its concreteness; the rational nature of the process only provided the form. "The sole thought which philosophy brings to the treatment of history is the simple concept of Reason: that Reason is the law of the world and that, therefore, in world history, things have come about rationally. . . . Only the study of world history itself can show that it has proceeded rationally, that it represents the rationally necessary course of the World Spirit. . . . This, as I said, must be the result of history. History itself must be taken as it is; we have to proceed historically, empirically" (*Reason in History*, pp. 11–12). Empirical research should thus be guided by the presupposition that the events exhibit rational meaning, unfolding a rational spirit. Hegel's own historical studies traced the growth of human freedom from the beginnings of the family to its culmination in the nineteenth-century state.

Hegel's direct influence was limited. His prose is cumbersome, his thought convoluted and abstract. But his indirect influence on modern thought has been deep. Marx, for example, by insisting on a material basis for the historical process and wrenching German idealism free from its abstract, spiritual preoccupations, laid the foundations for twentieth-century sociology. However much Western sociologists may dissent from Marx, they all share his notion that the various elements of society are interrelated in such a way that they can be analyzed and studied.

The capacity for analysis on which sociological method is based implies an objective separation between analyst and data. Kant made this separation possible by showing that the thinking self, in the process of understanding, acts in objective ways on social phenomena. Hegel gave this notion historical and sociological relevance by showing that the process itself of conscious activity was the basis of reality. In the words of Wilhelm Dilthey, who as much as any historian formulated a clear conception of the differences between the natural and social sciences, "the connections in the mind-affected world arise in the human subject and it is the effort of the mind to determine the systematic meaning of that world which links the individual logical processes involved to each other. Thus, on the one hand, the comprehending subject creates this mind-affected world and, on the other, tries to gain objective knowledge of it. Hence we face the problem how does the mental construction of the mind-affected world make knowledge of mind-affected reality possible?"[13]

Most modern historical writing is not philosophical. Hegel's philosophical history seems procrustean to practicing historians, despite his claims to the contrary. Nevertheless, Ranke's historical method, which seeks out primary facts through analysis of original sources, is as imbued with Kantian presuppositions as was Hegel's philosophical history. Ranke's own writing on history reveals his affinities to Kant. He spoke of himself as a Kantian and said that Kant had demonstrated how far our knowledge of things could extend.[14] In addition to his explicit acknowledgment of Kant's influence, Ranke's writings on history show that he accepted much of Kant's epistemology. In a passage written in the 1830s, Ranke, while accepting the traditional division between analytical and empirical knowledge, cautions against regarding history as simply a collection of historical facts.

> Nevertheless those historians are also mistaken who consider history simply an immense aggregate of particular facts, which it behooves one to commit to memory. Whence follows the practice of heaping particulars upon particulars, held together only by some general moral principle. I believe rather that the discipline of history—at its highest—is itself called upon, and is able, to lift itself in its own fashion from the investigation and observation of particulars to a universal view of events, to a knowledge of the objectively existing relatedness. (Stern, p. 9)

Ranke felt that the historian could discover the universal aspect of history by two methods. First, he should feel a "participation and pleasure in the particular for itself," and second, he should see the universal not in concepts but in the external relationships among the particular nations and individuals. Each of these methods presupposes Kant's picture of the act of knowledge. The historian can participate in the particular for itself only if particular events are considered not as noumena or things in themselves but as phenomena whose every aspect reflects the thinking subject organizing experience by its own a priori intuitions. Ranke's second idea, that the universal emerges when the interaction between nations and individuals is studied, depends on a sense of absolute time. Individuals and nations are interrelated because each can be located on an absolute temporal continuum which is itself without particular formal content. Universality can only be a necessary form of experience because of this temporal continuum. Otherwise the interconnection of particulars would imply preexistent concepts or moral values.[15]

These assumptions underlie Ranke's historical writings as well as his theoretical essays. In his *History of the Popes*, written in the same decade as the paragraph just quoted, Ranke attempted to explain how the papacy changed from an institution with universal spiritual claims into a secular Italian power. In a short passage early in the book he described the change in four paragraphs. In the first Ranke presented a general assessment of the change, commenting on the previous claims of the papacy and the increasing interest of the popes in political activity. The second paragraph quoted one of the speakers at the Council of Basel in the early fifteenth century. The speaker observed that he used to think that secular power should be separated from spiritual power, but that he had come to believe power without virtue was ridiculous. In the third paragraph Ranke quoted Lorenzo de' Medici, writing to Innocent VIII at the end of the fifteenth century. Lorenzo advised Innocent to bestow honors and benefits on his relatives, since his holiness and piety would bring him little reward. Ranke summed up in the fourth paragraph the significance of the development he had just illustrated. "There is an inner connection here in that the European powers were taking a part of the papal privileges at the same time that the pope himself was beginning to engage in secular activities."[16]

In this characteristic passage Ranke presented the objectively existing interconnection between two primary facts: the comment at Basel and the advice of Lorenzo. He did so by rising above the particulars to exhibit the universal process they were a part of. He showed the unity of this twofold change whereby the popes' spiritual pretensions declined as their political interests increased. The relatedness he suggested, however, depended on an important assumption. Both he and his readers had to assume a conception of absolute time giving an intrinsic meaning to the two events apart from any other quality. In his transition between the two primary facts Ranke acknowledged that they lacked any other connection than a purely temporal one. "In Italy they saw this [the issue of secular power] from another point of view at a somewhat later date." Ranke abstracted the meaning of these two events from the political, geographical, and personal categories which defined them as primary facts and presented their meaning as a pure consequence of their temporal relation. Not only that but the historian himself functioned as an absolute thinking subject, using his intuition of their temporal relation to create order out of the particular facts. Both the absolute time line and the absolute thinking subject underlie Ranke's narrative, giving it verisimilitude in the readers' minds, just as those assumptions lay behind his theoretical comments.

THE TWENTIETH-CENTURY CRITIQUE

The absolute time line and the unified thinking subject on which modern historical method is based are not simply items of blind faith. We expect them to describe our experience accurately and comprehensively. We cannot ignore persistent empirical data which these assumptions fail to explain. But how can we be sure they apply to all of experience? Though controlled experimentation in the social sciences can establish the validity of some types of hypotheses, only the natural scientists can isolate and interpret instances of noncompliance with this particular principle. The historian can never prove that specific data do not fit the time line. It is within the province of historical method to refine and clarify the temporal relatedness of the particulars. In Ranke's narrative of the growth of papal political interests, for instance, a contemporary scholar might ask whether there is evidence of an actual relationship between the view expressed at Basel and Lorenzo's later cynicism, might offer alternative hypotheses as to the meaning of the two quotations, or might even try to redate the events under consideration. But the scholar can never use the historical data to show that some historical events are not traceable to primary facts or that temporal sequence obscures important relationships.

By the same token, historians cannot ignore empirical evidence from the natural sciences that does not fit into Newtonian time. The very conviction of objectivity which underlies our claim to describe the past depends on our faith that all historical events can be reduced to primary facts capable of simple temporal location. Historians agree on no other way of verifying historical generalizations. If it is clear that some empirical data can not be fit into absolute time, then we must consider the possibility that primary source criticism, valuable as it is in its own sphere, is not a sufficiently comprehensive instrument to be used exclusively by historians in drawing up a representation of the past.

In the same decade in which Ranke died, an experiment by two physicists, Michelson and Morley, presented undeniable evidence of empirical phenomena that did not fit into absolute time. Attempting to measure the resistance of ether, they unexpectedly discovered that light traveled at the same speed whether moving in the direction of the earth's motion or across it. If the results of the experiment were correct, and successive repetitions in the last two decades of the nineteenth century left little doubt about that, then light did not fit within the framework of an inertial coordinate system. The constancy of the velocity of light brought into question the absoluteness of time and space, for an observer moving toward a light beam would find that it arrived at the same speed as if he were moving away from it. Of course the speed of light is so great that no experience in the everyday world could ever demonstrate the fact. The interferometer that Michelson and Morley used to produce this result is a complex instrument made possible by the great technological achievements of the nineteenth century, and it measured a difference that could not possibly be seen by the naked eye. But the mere existence of the data demanded some explanation; Newtonian physics could not offer one.

Within two decades other evidence accumulated that could not easily be fit into an absolute temporal sequence. In 1900 Max Planck argued that electrons move not in continuous paths but in discrete jumps, called quanta. While these jumps were predictable and depended on the quantity of heat, they still presented problems for Newtonian physics. At the subatomic level matter seemed to move in space without a temporal interval. Since Newton's concept of absolute space depended on absolute motion, this was a severe challenge to his conceptions of space and time. Here again nothing in everyday life corresponded to quantum motion, but Planck had shown for subatomic matter what Michelson and Morley had shown for astronomical distances. A single temporal dimension that could be measured without reference to spatial considerations did not incorporate all empirical phenomena.

Albert Einstein's answer to this problem is certainly one of the great achievements of the twentieth century. He began by identifying the two apparently incompatible postulates on which the science of his day was based. On the one hand all motion was measured by reference to a coordinate system which did not itself move and was therefore inertial. On the other hand the speed of light in a vacuum was absolutely constant. When the speed of light was measured against the inertial coordinate system associated with Newtonian physics, either it ceased to be constant or the system ceased to be inertial. Einstein argued that the system had to be changed. The three Cartesian coordinates of space could be reconciled with the uniform velocity of light only if the temporal coordinate were added.[17] We could no longer define matter by using the spatial coordinates to describe a fixed "now" and then relate this to other "nows" on a fixed temporal continuum. Since time and space were both affected by motion, questions of simultaneity, duration, and simple location were interdependent. Einstein's theory of relativity introduced no subjectivity or randomness into the universe; on the contrary it enlarged the field of predictability and established an absolute standard—the speed of light in a vacuum—to which all other motion was relative.[18]

Einstein did not concern himself much with the metaphysical implications of his theory. He thought the relationship between logical concepts and empirical data was a simple one. As he said in an essay on the problem of space, "concepts have reference to sensible experience, but they are never, in a logical sense, deducible from them. For this reason I have never been able to understand the quest of the a priori in the Kantian sense. In any ontological question, the only possible procedure is to seek out these characteristics in the complex of sense experiences to which the concepts refer" (*Essays*, p. 63).

But relativity theory, and the empirical data which lie behind it, are profoundly subversive of the assumptions whose development has been traced in this chapter. The concept of an ideally isolated system, based on the notions of simple location, substance, and the primary and secondary qualities, can no longer contain our experience. Alfred North Whitehead (1867–1947), more than another thinker, brought out these implications. He argued that the concept of an ideally isolated system leads to what he called the "fallacy of misplaced concreteness." Those succumbing to this fallacy identify the ultimate concrete fact with a bit of matter which can be simply located in space and time. Whitehead said this fallacy was pervasive in thought before the twentieth century, including not only the strictly Newtonian physicists but even Kant himself. Whitehead noted that Kant spoke in some passages of the infinite divisibility of matter and considered that the representation of the whole is necessarily preceded by its parts.

Against this long tradition of Western thought Whitehead proposed his philosophy of organism. The concrete fact ceased to be the simple object and became the event, whose reality was a process and which was called an event rather than a body because its time must be taken into account when describing it.[19] This approach had profound implications for our concept of time. Time remained, as it was in Kant's system, a process of synthetic realization that extended beyond the spatial and temporal limits of nature, but the necessity for a single series of linear succession was no longer present. Kant felt that the proposition "time has only one dimension" was necessarily true because it was synthetic and a priori, derived from the essential structure of the human mind. Whitehead disagreed, arguing that our consciousness revealed a more complex temporal relationship in experience. As evidence that time should be conceived as a group of linear serial processes, Whitehead appealed "(1) to the immediate presentation through the sense of an extended universe beyond ourselves and simultaneous with ourselves, (2) to the intellectual apprehension of a meaning to the question which asks what is now immediately happening in regions beyond the cognisance of our senses, (3) to the analysis of what is involved in the endurance of emergent objects" (Science, p. 124).

Whitehead thus argued that our immediate experience of events, our intellectual awareness of events outside this immediate experience, and the phenomenon of change itself are all events whose reality lies in the process of emergence into space-time. Their sense of simultaneity depends on this emergence and not on their ability to be located within a single time frame. The insight that each event has its own definite meaning of simultaneity is not dependent on relativity theory. (Einstein in fact rejected the notion that there was more than one linear time series.) It depends on overcoming the fallacies of misplaced concreteness and simple location, on seeing the process rather than the bit of matter as the most concrete entity (Science, p. 126).

After Einstein and Whitehead the idea of simple location, which underlies modern historical method, no longer had the status that it did in the eighteenth century. The two intuitions of space and time which for Kant gave apodictic certainty to mathematical judgments no longer encompassed all available data unless they were joined into a single intuition. Furthermore an analysis of our own experience using the concept of space-time leads us to the possibility of a group of time series each with its own meaning of simultaneity. These challenges to simple location do not mean that historians must give up the methods of source criticism which are founded on the Kantian framework. They do mean that we must consider the possibility that there are events of temporal significance—and hence of historical relevance—that cannot be simply located as primary facts on a single temporal continuum. To admit the possibility of such facts is not to give history over to the realm of fantasy; it is simply to look for other means of convincing readers of the truth of the narrative.

Relativity theory, although it undermined the notion of absolute time, did not in any serious way attack the absolute thinking subject as the basis for objectivity. Einstein and Russell both went out of their way to defend the observer's capacity to make unified judgments, given adequate coordinates whose meaning is fully understood. Whitehead, more clearly aware of the philosophical issues, realized that his philosophy of organism made the subject part of a field of mutual interactivity. He even substituted the word "prehension" for the traditional word "perception" to emphasize the mutual character of the act of knowledge. Acknowledging that the Cartesian subject was no longer an adequate premise, he maintained that "no individual subject can have independent reality, since it is a prehension of limited aspects of subjects other than itself" (*Science*, p. 151).

The decisive scientific challenge in the twentieth century to the absolute subject, and hence to the grounds of objectivity, comes from the field of psychoanalytic theory. To understand this critique we must consider again how Hegel modified Kant's idealism. The German idealists who followed Kant, by offering to describe noumenal reality, gave to their thought a practical appeal that his lacked. Kant's insistence that we can know nothing of things in themselves proved to be the least palatable part of his system. By claiming to talk about the real world, even though the world was conceived as a manifestation of consciousness, Hegel brought new force and vigor to the rationalism of his predecessor.

But the claim to discuss noumenal reality brought risks. Kant's division had protected the noumenal subject from questions that applied only in the phenomenal world. In particular he showed that the subject did not need existential unity. Reason demanded only a logical unity to provide a satisfactory basis for syllogistic reasoning. Hegel identified the logical unity of the subject with existential unity. To be sure, this unity was one of process and not of unchanging traits or categories, but it sufficed to give direction and meaning to the temporal processes which exhibited the action of reason in history.

Furthermore the unified subject had to possess some sort of existence which transcended the temporal process. This existence could never be clearly stated, since all human experience was temporal, but the subject nonetheless existed as a conceptual point outside of time, giving perspective to all events in time. Hegel, for example, and other German idealists used the term "absolute"; Marx gave the absolute a material basis and called its social manifestation "proletarian communism." In both cases these absolute, timeless concepts gave the processes of time itself a teleological dimension. They further integrated the subject into the temporal process. For Hegel time was real partly because it was the product of a thinking, purposive subject, and it expressed in a partial and limited way that purpose. For Marx the process was still unified by a goal, though it did not have the rational character of Hegel's absolute.

The unity of the subject, then, though not its rationality, was a precondition of objectivity. In the same decade as the Michelson-Morley experiments, Freud began a process which was to call into question this very premise. Freud's studies with Breuer and Charcot in the 1880s convinced him that many physical symptoms had psychological causes that could be treated directly instead of through physical manipulation of presumed neurological deformations. Freud's clinical goal of treatment led him to map out the structure and essential processes of the human mind. Freud built his system on two basic assertions, both of which reflect the achievements of Kant and Hegel.

In his first assertion, that mental processes are largely unconscious, Freud was explicitly aware of his debt to Kant.[20]

> In psychoanalysis there is no choice for us but to assert that mental processes are in themselves unconscious, and to liken the perception of them by means of consciousness to the perception of the external world by means of the sense-organs. . . . The psychoanalytic assumption of unconscious mental activity appears to us . . . as an extension of the corrections undertaken by Kant of our views on external perception. Just as Kant warned us not to overlook the fact that our perceptions are subjectively conditioned and must not be regarded as identical with what is perceived though unknowable, so psychoanalysis warns us not to equate perceptions by means of consciousness with the unconscious mental processes which are their object. Like the physical, the psychical is not necessarily in reality what it appears to us to be. We shall be glad to learn, however, that the correction of internal perception will turn out not to offer such great difficulties as the correction of external perception—that internal objects are less unknowable than the external world.

In the last sentence Freud shows the confidence of post-Kantian thinkers that they could know noumenal reality as long as they kept in mind Kant's insistence that it always be related to the perceiving subject.

If Freud's first premise revealed the influence of Kant, his second reflected a dialectical cast of mind with strong affinities to Hegel. He asserted that sexual impulses play a primary role not only in the causation of mental disease but in the cultural and social achievements of mankind.[21] This assertion involved a redefinition of sexuality, which became in Freud's writings no longer a statically conceived trait but a dialectical process in which different objects and activities were sexual at different stages of human life. Adult sexuality depended on the manner in which objects had been presented for gratification at an earlier stage. He said that people had conventionally defined sex in vague terms as "something which combines a reference to the contrast between the sexes, to the search for pleasure, to the reproductive function and to the characteristic of something that is improper and must be kept secret" (*Lectures*, 20; *Works*, 16:304). To clarify this vague collection of notions, he went on to analyze infant sexuality, noting the development from oral to anal to genital phases, discovering a variety of components in the sexual instinct dialectically related to produce the full complement of cultural values and libidinal drives of the adult person (*Lectures*, 21; *Works*, 16:320–38).

In constructing this picture of adult sexuality Freud assumed that the development he described was governed by a single pleasure principle. By 1919 his own experience had disabused him of that hypothesis. In the wake of World War I many of his patients experienced traumatic dreams in which they relived the bombings, trench warfare, or other horrifying events they had experienced. Given Freud's interpretation of dreams as wish fulfillment, these traumatic dreams were difficult to understand. Shaken by these phenomena which did not fit his theory of dreams, Freud began to find puzzling certain features of ordinary children's play. Childhood behavior was important to Freud, since it reflected instinctual expression with a minimum of social and conscious interference. He observed a friend's child repeatedly throwing away his toys. What puzzled Freud was the evident pleasure the child felt from the act. After lengthy observation Freud confirmed that the child derived as much satisfaction from throwing the toy away as in getting it back again; in fact the pleasure at throwing away objects seemed to be not at all dependent on the thought that they would return. Freud concluded that even in childhood painful experiences could be felt as pleasurable.[22]

As a final challenge to his previous hypothesis of a single pleasure principle, the resistance of his patients to treatment became a cause of increasing concern. Freud had come to associate this resistance with the regression that he felt lay at the heart of neurosis. Since regression involved a compulsion to repeat an earlier stage of development, a stage which had not produced the satisfaction appropriate to it, he had again discovered a phenomenon inexplicable if one assumes a unified pleasure principle. How could a patient be determined to repeat an activity which produced no pleasure? Why was the patient resistant to the psychoanalyst's explanations of the causes for this compulsion to repeat? By 1919 Freud had come to the hypothesis that these questions could not be answered within a clinical context. They were fundamentally conceptual in nature and required a major reformulation of the basic notion of pleasure.

In *Beyond the Pleasure Principle* Freud attempted just such a reformulation. He began by analyzing conscious pleasure in quantitative terms. "We have decided to relate pleasure and unpleasure to the quantity of excitation that is present in the mind but is not in any way 'bound' and to relate them in such a manner that unpleasure corresponds to an increase in the quantity of excitation and pleasure to a diminution. . . . The factor that determines the feeling is probably the amount of increase or diminution in the quantity of excitation in a given period of time" (*Beyond*, chap. 1; *Works*, 18:7–8). This conception of instincts goes back to Freud's earliest training as a physiologist and reflects the mechanistic picture of the universe which dominated European thought from Descartes to Einstein. When he studied physiology under Brucke in the early 1880s, Freud was especially impressed by his teacher's definition of organisms as phenomena of the material world, systems of atoms, moved by forces of repulsion and attraction.[23]

This definition of physiology serves well to explain conscious pleasure but it does not explain unconscious pleasure, for Freud had learned from Kant that the world of the unconscious lies outside the reach of time.

> As a result of certain psychoanalytic discoveries, we are today in a position to embark on a discussion of the Kantian theorem that time and space are 'necessary forms of thought'. We have learnt that unconscious mental processes are in themselves 'timeless'. This means in the first place that they are not ordered temporally, that time does not change them in any way and that the idea of time cannot be applied to them. These are negative characteristics which can only be clearly understood if a comparison is made with conscious mental processes. On the other hand, our abstract idea of time seems to be wholly derived from the method of working of the system [of conscious perception] and to correspond to a perception on its own part of that method of working. (*Beyond*, chap. 4; *Works*, 18:28)

Since the instincts must exist in the unconscious without reference to time, they cannot be reconciled dialectically; they remain two separate and contradictory impulses. One of these instincts derives satisfaction from the release of tension; it is conservative, seeking to restore an earlier state of things, and is ultimately an instinct to return to an inorganic state. Freud called this the death instinct, or Thanatos, meaning by it not a conscious wish to die, but an unconscious striving to reduce tension. The second instinct is creative, seeking to build larger and larger unities. Freud called it Eros. It is, he said, the true life instinct, associated with the sexual instincts which seek by the processes of reproduction to incorporate others into new unities (*Beyond*, chap. 7; *Works*, 18:34–43). At the unconscious level, then, the instinctual drives of the individual are deeply ambivalent, made up of impulses which are polar opposites and mutually frustrating. Consciously we feel these drives as a single process but only because our conscious perception is time-bound. At the unconscious level considerations of time cannot serve to reconcile the opposing instincts.

Freud applied this insight on both the personal and social levels. In *Beyond the Pleasure Principle* he concluded by analyzing the orgasm. "We have all experienced how the greatest pleasure attainable by us, that of the sexual act, is associated with a momentary extinction of a highly intensified excitation" (*Beyond*, chap. 7; *Works*, 18:62). Thus even in the most encompassing individual pleasure there are elements of tension and release, perceived as conscious pleasure but analyzable into contradictory unconscious instincts. The unconscious experiences frustration even there.

In later works, particularly *Civilization and Its Discontents*, Freud extended the struggle between Eros and Thanatos to include all of human history. "And now, I think, the meaning of the evolution of civilization is no longer obscure to us. It must present the struggle between Eros and Death, between the instinct of life and the instinct of destruction, as it works itself out in the human species. This struggle is what all life essentially consists of, and the evolution of civilization may therefore be simply described as the struggle for life of the human species." [24] In applying his insight to both social and individual history, Freud offered a major alternative to Descartes' separation of subject and object. He showed that the individual's particular acts of judgment and decision making could not in fact be isolated from their historical background, that the values of an individual's society reflected the same instinctual ambivalence that lay at the heart of personal drives. In its attack on the unity of the absolute subject, Freud's analysis of instinct represents a challenge to the concepts which underlie modern historical method as deep and unavoidable as that represented by the substitution of relative for absolute time. [25]

Freud and Einstein do not, to be sure, offer the only challenges to Kantian metaphysics. As early as 1854 Georg Rieman suggested non-Euclidean geometries, in which the traditional Cartesian coordinates were no longer applicable. Wittgenstein, after an early career in which he developed a logic based on the premise that all meaningful statements were statements of fact, abandoned this logical positivism in the 1930s. In his later period he argued that logic was only one of many language games in which words can be not only names of objects but anything else and where language is no longer a picture of reality but a tool with many uses. Gödel in the same decade even called into question the status of mathematics itself as a self-contained and comprehensive system. He showed that any mathematical system will be either incomplete or inconsistent; it will produce true statements which cannot be deduced from its original assumptions. In 1928 Werner Heisenberg showed that the normal intuitive concepts of time, space, and causation did not apply at the subatomic level. J. S. Bell suggested in 1964 that the fundamental uncertainties that had been recognized in quantum mechanics at the subatomic level since Heisenberg applied also to the ordinary world.[26]

It may seem unfair to stress the implications for history of relativity theory and of quantum mechanics. Certainly everyday life can be conducted without reference to the space time continuum. Why cannot historical writing afford the same luxury? We can indeed live from day to day without using relativity theory, but by the same token we can also lead our daily lives without Newton. Aristotelian science is more than adequate for most quotidian experience. When asked to describe the contents of a room, most of us would say, "that chair is three feet from the wall and the lamp is standing just behind it." Few of us would create a system with three coordinates intersecting at an imaginary point in order to describe our surroundings. Our immediate reaction when asked to date our first sexual experience would probably be to use our age or our stage in school. Only secondarily would an absolute date come to mind.

The fact that Newtonian time and space are not particularly useful in our personal lives has not prevented them from producing a technology which has markedly changed our lives and which does depend on absolute space and time. We may think of our telephone as sitting on the desk, but the satellite which allows us to make long-distance calls was launched in absolute space and time. In addition to this technology Newtonian time produced a remarkable historical method which opened up whole new realms of historical facts to critical evaluation. Relativity theory has yet to produce such a pervasive technology, but it has certainly begun to affect our lives in important ways. Nor is its importance diminished by the fact that we continue to describe our experience in more traditional ways.

We must be alive to the likelihood that Einstein will eventually change our concept of historical time as much as did Newton. That relativity theory is a minor part of our daily lives does not mean it poses no challenge to historical method. In fact this challenge cannot be dismissed as purely metaphysical. The assumption of absolute time tends to exclude from historical relevance any events which have no potential for simple location on a time line. The realm of unconscious motives, for example, where earlier events are determining present conduct through a process not precisely traceable in time, is thus theoretically excluded from history. Even more untestable are the underlying cultural attitudes and archetypes which emerge from a shifting series of concrete events but which cannot be traced to any simple temporal order. Finally the deepest economic facts of society, the hidden social relations which determine what sorts of labor are compensated in what way, cannot be precisely dated, for they are hidden in the tangible commodities.

Faced with generalizations of this nature most historians recoil in horror. Because none of these generalizations are testable by reference to primary facts that can be simply located, historians tend to assume they are exercises in fantasy. Yet such phenomena are part of the past, just as they are part of everyday life. A historical method that excludes them is questionable. Such a method is especially questionable if it cannot be justified by rational arguments and empirical evidence. In this case the assumptions about time that exclude these phenomena also exclude some data which are so plainly evident to contemporary scientists that they have devised a new sort of time.

Abandonment of simple location as the test of the primary fact does not give history over to uncontrolled fantasy. In recognizing the limitations of Newtonian time, historians need not give up the tried and true techniques of source criticism that the idea of simple location has produced, any more than we have to give up satellites or computers. In the sixteenth century, when Europe adopted the solar system, sailors did not have to give up navigational techniques premised on the fixed earth. New technologies and new critical techniques can be added to old ones; they need not replace the earlier ones completely.

Nor does the challenge to objectivity posed by the elimination of the unified subject necessarily lead to propaganda and undisciplined historical writing. For historians to acknowledge that they are part of the phenomena they record can serve to make them more aware of the values and preconceptions they bring to their work. Such an awareness can increase the clarity with which they communicate these values and sharpen the significance of their writing.

These are not simply theoretical possibilities, for the historians who dominated Western historiography from Herodotus to Guicciardini approached their subjects without assuming either an absolute time line or an absolute subject. They felt neither that every historical fact was capable of simple location nor that they existed as unified subjects separated from their inquiry. Yet they produced works which formed major elements of Western social, political, cultural, and personal identity, while offering to their contemporaries a study of the past which was useful and inspiring. How they accomplished these goals through their own conceptions and measures of time will be the subject of the following chapters.

3

The Relative Time of Herodotus and Thucydides

For fourteen years the thirty years' truce which had been concluded after the capture of Euboea remained unbroken; but in the fifteenth year, when Chrysis was in the forty-eighth year of her priesthood at Argos, and Aenesias was ephor at Sparta, and Pythodorus had still four months to serve as archon at Athens, in the sixteenth month after the battle of Potidaea, at the opening of the spring, some Thebans . . . about the first watch of the night entered under arms into Plataea.

Thucydides, *History of the Peloponnesian War*, 2.2

THE MEASURE OF RELATIVE TIME

We cannot hope to understand past chronologies until we overcome the tendency to look on them as primitive versions of our own, remarking their clumsiness and vagueness but not seeing the functions they served for the historians who used them. The chronology Thucydides used to locate the beginning of the Peloponnesian war offers an excellent example of the sort of dating that seems to justify this condescension. Where Thucydides had to date his war with reference to several polities as well as the time of year, we can express the date simply and economically with a single number that transcends all particular states and events and that constitutes an absolute date.

Certainly from the perspective of absolute time, condescension toward past chronologies is fully justified. The development of absolute chronology since the seventeenth century has brought about a remarkably complete and accurate picture of those past events that can be placed on an absolute time line. Furthermore, this achievement has depended in large part on seeing clearly the deficiencies in past systems. The great chronologers of the modern period have devoted themselves to studying the points of vagueness in the systems which major historians of the past used to give temporal organization to their events. Yet their very interest in precise dating is itself part of the conceptual developments traced in the preceding chapter. As François Chatelet, one of the leading students of historiography has remarked, "The necessity for an exact determination of chronology appears only at the moment when we have both a concept of time that renders it indispensable and the intellectual tools that permit it to be practiced effectively. For the recognition of human historical consciousness, the conception of time and the techniques of historical discourse can only with difficulty be considered separately."[1]

The historians who wrote before Newton did not have such a concept of time as Chatelet describes. They saw times rather than time; they accepted the existence of different temporal sequences and relations in their stories. Virginia Hunter, in an insightful study of Herodotus and Thucydides, has said of Thucydides' sense of time, "In the past he saw not a straight line but rise and decline, process and the repetition of process, the important stages of which, growth, achievement, peak and regress, he described. Also included in his history of the distant past were great gaps of time, which he did not attempt to explain or fill, since he accepted discontinuity."[2] Thucydides created a linear series for the war, and once he had located its beginning he dated its events clearly by summers and winters from one year to the next, but other historical realities did not fit easily within this linear scheme, realities which were important to understanding the causes and outcome of the conflict. Similarly Herodotus, though he gave some aspects of his story of the Persian wars a linear dimension, preferred to place most of the events within another sort of time. While it is undoubtedly true that these historians could have dated their stories more precisely with the modern dating system, it is not true that they would have preferred it or would have seen in that system a greater truthfulness than their own.

Herodotus and Thucydides used their dating systems to express two fundamentally different sorts of temporal relationships. Some of these relationships were linear, where the temporal order of antecedent events had a determining influence in shaping the final result; others were episodic, where the events had a temporal dimension but where the chronological order of specific events within an episode had no particular meaning and where one event in the episode could not be treated as the efficient cause of another. Episodic time was discontinuous, emphasizing process rather than progressive building of events on one another. Its results could be cumulative, but the exact temporal order was not an important factor in the process that produced the final effect. A single absolute chronology could not express accurately these two elements of time. The Greeks needed other means; their techniques may seem primitive to us, since they cannot be easily used to create a linear series of simply located events, but in fact they conveyed temporal dimensions which our commitment to an absolute time line makes hard to express in modern historiography.

When modern readers encounter Herodotus' account of the Persian wars, they often see first the famous events that have formed such a vital part of Western pride—the Athenian victory at Marathon, the valiant Spartan defense against the Persian hordes at Thermopylae, the defeat of the Persian navy at Salamis. When Herodotus described the conflict between Persia and Greece, he chose not to see the war itself as the primary event but to look beyond the war to the deeper issues which brought it to being. He presented the war as a conflict between the Greeks and the barbarians, a struggle between Europe and Asia arising out of fundamental disagreements about values and social order. For example, the two sides had radically different perceptions of the importance of women. Asians could not understand why the Greeks attacked Troy to recover the abducted Helen. "We think . . . that it is wrong to carry women off: but to be zealous to avenge the rape is foolish. . . . We of Asia regarded the rape of our women not at all; but the Greeks, all for the sake of a Lacedaemonian woman, mustered a great host, came to Asia, and destroyed the power of Priam." [3]

In focusing on these cultural issues Herodotus addressed a class of realities which lay between the timeless and abstract concepts of philosophy and those historical events which could be unequivocally located on an absolute time line. Europe and Asia were events rather than abstract concepts to Herodotus; they contained elements of geography, ethnicity, culture, society, and politics, but they could not be comprehended within any of these modern categories. The differences between the two could be traced back to the distant past, before the Trojan war. Greece and Asia constitute the basic temporal building blocks of the narrative. Herodotus used concrete terms like "Persians," "Greece," or "peoples" to describe these realities, but the terms he used had a meaning that went beyond the physical boundaries and the inhabitants. They are not timeless entities, possessed of innate characteristics, like triangles or virtues, but subtle collections of attitudes, political practices, and social values which led the two sides into conflict. Unlike abstractions, they occupy space and time. "The Persians claim Asia for their own, and the foreign nations that dwell in it; Europe and the Greek race they hold to be separate from them" (*Histories*, bk. 1, chap. 4).

To describe the temporal aspects of the conflict Herodotus created two fundamentally different chronologies. On the one hand he used several episodic chronologies that measured time by generations and that were tied to the spatial dimensions of the conflict. Herodotus oriented his reader by noting the reigns of both the Asian and the Greek rulers. These reigns expressed relations that were temporal but not linear. They measured duration, but the episodes recounted under a particular reign had no specific temporal location within the reign. Modern scholars have often expressed frustration at this practice and have tried to make a linear time line out of Herodotus' narrative by discovering how long he thought a generation was. They have put the figure variously at 23, 26, 32, 34, 39, and 40 years, but a recent student of the subject has contented himself with a simple expression of confusion.[4]

In fact the generations have no quantitative aspect. They exist as pure indications of the fact of duration; the relation among separate generations is discontinuous and extrinsic. But Herodotus was not indifferent to progressive and continuous elements of time. Alongside this episodic chronology he also drew up a linear sequence of years leading back to the dimmest recesses of known time. By this second chronology he conveyed the temporal dimension of Greek culture in a linear fashion. These two chronologies are fundamentally different in their orientation and function, though they combine to give Herodotus' narrative a richness and subtlety it would otherwise lack. By examining the episodic and linear chronologies in turn we can see more clearly the separate functions they served in Herodotus' work.

Though Herodotus used it to present important moral and political lessons that resonate through the story, the episodic chronology based on the reigns of the kings derived its meaning from the spatial boundaries of the kingdoms. Only after describing the territory in Asia Minor that King Croesus had ruled did Herodotus establish the chronology of the kingdom, noting that the Heraclid family, which Croesus' succeeded, had reigned twenty-two generations, a period of five hundred and five years. The numbers may seem reassuring, but where does that figure locate the events? The reader is given no indication, because the absolute location is irrelevant to the episodic time Herodotus described here. The five hundred and five years serve only to show the reader that the family was well-established as rulers of Lydia when the last Heraclid, Candaules, lost his throne and Croesus' ancestor came to rule. The years do not measure specific events, they constitute a period whose length alone is important, for they had given the Heraclids confidence in their possession of Lydia.

This confidence and the foolhardy behavior it inspired are among Herodotus' major themes. They explain much of the behavior of all the Asian kings and particularly of Darius and Xerxes, whose attack on Greece constitutes the focus of Herodotus' story. Here at the beginning of his narrative, immediately after telling us of the length of the Heraclids' reign, Herodotus introduces these themes with the story of Candaules, the last of the Heraclid line and the inheritor of so much arrogance and assurance. "This Candaules, then, fell in love with his own wife, so much that he supposed her to be by far the fairest woman in the world; and being persuaded of this, he raved of her beauty to Gyges, son of Dascylus, who was his favorite among his bodyguard; for it was to Gyges that he entrusted all his weightiest secrets" (1.8). Abruptly, with this sentence we find ourselves in the court of Candaules, as he persuades a reluctant Gyges to hide behind his bedroom door and watch the queen undress. Catching a glimpse of Gyges and indignant at her husband's behavior, she forces the distressed bodyguard to help her overthrow Candaules. The royal whim has brought to an end half a millennium of rule.

The episode of Gyges and Candaules does not emerge from a specific set of antecedent conditions; but neither is it simply the illustration of an abstract principle. It takes on temporal dimensions by standing against the background of the long duration of the Heraclids' reign. Herodotus saw forces at work in the episode that could not be comprehended in a linear time frame, for the downfall of Candaules fascinated him as an example of *metabole*, dramatic and sudden reversals of fortune. He felt they were important, meaningful, and in some sense predictable, but not simply a product of a specific series of actions in the past. Instead he saw an overarching fate which determined these acts of arrogance and led the leaders of Asia to destroy themselves. He expressed this aspect of the story of Candaules when he observed that the king made the indecent proposal to Gyges out of some vague necessity to end badly.[5]

Duration was not the only temporal factor that Herodotus measured with the generational dating system. The reigns he used to measure time are themselves temporal units whose place in a series can be more significant than their absolute length. In the case of Croesus what is important is not the number of years he reigned, but that he was the fifth of his family to rule Lydia. This place in the order of kings doomed him to lose his throne, for when Gyges took the throne an oracle told him that his family would last five generations.

Events within Croesus' reign do not have a meaningful intrinsic chronological relationship; they take their order from Herodotus' thematic interests. Early in the account Herodotus told of Croesus' scorn for Solon's admonition that he should not allow his wealth to make him overconfident of permanent happiness. The traveling Athenian's warning, presaging the metabole that afflicted Croesus' fortune when he lost his empire to the Persian king Cyrus, is a dominant motif of Croesus' reign. It takes precedence over the historical events that brought about his fall. As the condemned king lay on the pyre awaiting execution, he thought back not to the generations of rulers that had preceded him, not to the act of hubris that had brought his own family to power, not even to the prediction that the Heraclids would have their revenge after five generations. None of these antecedent conditions was present in his consciousness. Instead he remembered Solon's admonition, a part of his own experience, to be sure, but a communication from outside his historical ambiance (1.86).

Solon's visit has only the loosest temporal relationship with other events in Croesus' reign. Herodotus never told exactly when the Athenian came to the court, nor how his warning related to other events. Those parts of his reign that Herodotus told us about reinforced the general picture of Croesus as a man determined to ignore the warnings of the fate that waited for him. He misinterpreted the words of the Delphic Oracle that a mighty empire would fall if he went to war with Cyrus. He ignored another oracle warning him of the death of his son. When the son died as a result of his defiance, Croesus did not think back to the words of Solon, though they were closer to him in time than they were on the pyre. Instead he "sat down in sorrow for two years" (1.46).

Herodotus' interest in presenting moral lessons through accumulation of temporal events rather than through a sequence of progressive steps is equally pronounced in his treatment of the Arachmenid rulers of Persia. Though he devoted most of the work to them and lavished a wealth of detailed anecdotes on their reigns, chronological indications are much the same as they were with the Lydian kings. Occasionally he noted the length of a siege if there was some tactical significance to its duration,[6] but normally the length of the reign is the only measure of time. Persian kings, like the Lydians and Heraclids, fell to excesses of daring and pride. Cyrus overreached the strength of his army in fighting the Massagetes, was caught on the far side of the river dividing the two regions, and was killed by Queen Tomyris. His son Cambyses went mad in Egypt after defying the religious customs of that country. Darius and Xerxes ignored a variety of signs and arguments which should have dissuaded them from their expeditions against Greece. Herodotus assembled these stories to support his contention that an irresistible fate directed the self-defeating behavior of these kings.

But these excesses have no background. The separate acts of hubris are thematically related but do not build on one another. None of these kings thought of the pride of previous rulers; none fell through mistakes others had made in the past. Herodotus did not mean his narrative of these reigns to depend on previous events. The reigns are separate, self-contained units filled with episodes, not parts of a series chronologically related to one another. They have a temporal significance but no linear pattern.

Because he tied each time frame to a particular space, Herodotus ignored opportunities to synchronize events occurring in different countries, maintaining separate chronologies where a modern historian would create a single one. He could, for instance, have easily integrated the Lydian and Mesopotamian regions into a single dating system. When discussing Croesus' reasons for provoking Cyrus to war, he noted that Astyages, the Median king whom Cyrus had overthrown, was Croesus' brother-in-law and that his overthrow had been used as a pretext for attacking Cyrus (1.73). This intersection of Median, Lydian, and Persian dynasties could have afforded a general reference for integrating the three kingdoms' histories, but Herodotus was interested only in the events which bore directly on Croesus at the moment in the narrative where he mentioned the Median connection. There, after telling the story of how Croesus came to be Astyages' brother-in-law, he passed on to other things, promising to tell the reader more about Cyrus later. When, in a subsequent passage, he introduced the generations of kings in the Mesopotamian region, he made no reference to this earlier passage (1.130). The two chronologies remained without connection, separated in time as they were separated in space.

As the participants in Herodotus' story moved from place to place, they left their own times behind and acquired the times of the places they entered. This appears dramatically as Greek and Asian affairs interacted. Greece first entered the narrative when the Delphic Oracle advised Croesus to form an alliance with the most powerful Greek states. His inquiries revealed that there were two powers, the Athenians among the Ionian peoples and the Spartans among the Dorians. At this point Herodotus introduced the two cities, following his general practice of identifying the rulers in power at the time of Croesus' inquiries—Pisistratus at Athens and Leon at Sparta. After brief anecdotes to characterize these rulers, Herodotus returned to the fate of Croesus, leaving the Greek cities in their own time.

Though he established a chronology for Greece in this passage, other Greeks who entered the narrative in Book One were not connected with it. They seemed to carry no dates with them as they moved in and out of the regions of Asia. Solon, for instance, passed through Lydia shortly before Croesus made these inquiries, prompting Herodotus to mention that he was traveling in order to be away from Athens for ten years after he had given the Athenians their laws (1.29). Herodotus did not connect the laws of Solon with Pisistratus' tyranny, though he had already told his reader of that period of Greek history, and the laws and the tyranny must have had a close temporal relation. Modern historians, trying to create a linear order in Greek political history, have used the events Herodotus described, but they cannot date them precisely, since Solon's laws and subsequent travels may have occurred in the 590s or the 570s.[7] The matter cannot be settled by reference to Herodotus, since he considered Pisistratus and Solon to be part of two different and unrelated episodes. Solon's ten years were part of the episode by which Croesus came to understand the limits of his wealth and power, while Pisistratus' tyranny was part of the process of Greek political history. Having no thematic connection, they had no temporal relationship in Herodotus' mind.

Travelers in the other direction, from Asia to Europe, passed out of Asian times and into Greek ones. In Book Five, Aristagoras of Miletus, the Ionian whose revolt precipitated Darius' attack on Greece, traveled to Greece to ask help against the Persians. As he reached Sparta, he entered Spartan time. "At Sparta, Anaxandrides the son of Leon, who had been king, was now no longer alive but was dead, and Cleomenes son of Anaxandrides held the royal power" (5.39) Before telling of the Ionian's mission, Herodotus explained how Cleomenes had become the Spartan ruler. As Aristagoras moved on to Athens, he entered Athenian time. Here again Herodotus filled in the chronology, noting the succession from Pisistratus and promising, "All the noteworthy things that they did or endured after they were freed and before Ionia revolted from Darius and Aristagoras of Miletus came to Athens to ask help of its people—these first I will now declare" (5.65).

Herodotus had reasons beyond the simple desire for completeness for introducing the Athenian and Spartan chronologies here. He saw thematic significance to the events, just as he had with the genealogies of Asian kings. The episodes in Asian time showed the effects of long-lasting power in creating hubristic behavior; the episodes in Greek time showed the political wisdom of the Greek cities, which stemmed from the participation of the citizens in government. This theme found full expression in the later sections of the work as a background for the lengthy debates over strategy in the face of Xerxes' invasion. In the early sections the episodes introduced the theme. In the introductory passage on Sparta, for instance, Herodotus told how Sparta came to take over Tegea through the cleverness of one of its citizens in figuring out the meaning of an obscure prophecy (1.67–68).

Herodotus did not see this political wisdom as an innate and timeless characteristics of the Greeks. It resulted from events with spatial and temporal location. "Before this [the kingship of Leon and Hegesicles] the Spartans were the worst governed of well nigh all the Greeks. . . . Thus then they changed their laws for the better" (1.65). Not only did Greek states gain wisdom through specific events, but Herodotus used this long background of historically acquired wisdom to criticize the veracity of other specific events. He doubted the story that Pisistratus had come to power by a silly trick involving a giant statue of Athena. "[This plan] was so exceeding foolish that it is strange (seeing that from old times the Hellenic has ever been distinguished from the foreign stock by its greater cleverness and its freedom from silly foolishness) that these men should devise such a plan to deceive Athenians, said to be the cunningest of the Greeks" (1.60).

The subordination of the episodic chronologies to spatial orientation is clearest as the major movement in the Persian wars occurred, the passage of the great army of Xerxes into Greece. Herodotus followed the building of the army and the decision to invade Greece from Xerxes' accession to the throne, but as the army crossed the Hellespont it entered a new time frame. Without any attempt to synchronize the invasion with the reign of Xerxes, Herodotus calculated the arrival of the army in Athens according to the list of archons in that city. Treating the arrival in Europe as the beginning of a new time, he noted, "Now after the crossing of the Hellespont whence they began their march, the foreigners had spent one month in their passage into Europe, and in three more months they arrived in Attica, Calliades being then Archon at Athens" (8.51). The list of archons extended backwards at least to the period before Pisistratus, but even with Greek events Herodotus had no interest in forming a single time sequence, for he had made no reference to the list when introducing Athenian political life in the earlier books.

These episodic chronologies expressed events whose linear sequence was not an essential part of their meaning. They derived their significance not from antecedent events but from general themes which emerged in the narrative. By conveying these themes in a narrative with temporal order Herodotus removed them from the realm of abstract theory and made them real events. They occurred in time but were not wholly comprehensible by analysis of the linear sequence of earlier events.

Beside these episodic chronologies Herodotus used another in his work, one which measured linear time and provided a historical context for the themes narrated in episodic time. In Book Two he introduced an Egyptian sequence whose inaccuracies and contradictions have bothered generations of modern historians and played an important role in the establishment of the absolute time line itself. The problems inherent in dating the events of this book have prompted the conclusion either that Herodotus was simply in error or that the papyrus rolls he used to organize the lists of pharaohs had been disordered before he consulted them.[8]

But the confusion in Herodotus' Egyptian chronology applies only to the political succession of the pharaohs, and it is not a political sequence that Herodotus established in Book Two. Instead he created a linear sequence that would help explain the cultural differences that underlay the conflict between Asia and Europe and formed the thematic unity of the episodic chronologies. Herodotus connected Egyptian and Greek history not by integrating the political events in the two countries but by comparing the relative antiquity of their respective gods and by arguing that the Greeks took their gods from Egypt. To create this linear sequence in which Greeks and Egyptians had learned their religious practices and to trace the distinctive quality of Greek religion to specific antecedent events, Herodotus adopted the position that Egypt had a beginning in time and that Egyptian culture was the result of a linear and nonrepeating process founded in deep natural and physical changes.

The difference between the chronology in Book Two and the other chronologies in the work is immediately apparent. After ending Book One with the death of Cyrus, Herodotus began the next book with his son Cambyses' expedition against Egypt. Before describing that expedition, he introduced his readers to the affairs of Egypt. "Now before Psammetichus became king of Egypt, the Egyptians deemed themselves to be the oldest nation on earth" (2.2). This introduction is uncharacteristic. In beginning the chronologies of Book One, Herodotus always mentioned the monarch reigning at the point of time the story first encountered a particular region—first entered its spatial/temporal coordinates—and then moved back from the point of entry to the origin of that monarch's family. Psammetichus was not the monarch reigning when Cambyses invaded Egypt, nor was he the first of the pharaohs. Much later we are informed that he was the first in the line leading to Amasis, the reigning pharaoh in the time of Cambyses. But that connection did not concern Herodotus here. He introduced Amasis in customary fashion only at the beginning of Book Three, where we finally discover that it was against Amasis that Cambyses led his army.

In Book Two, Herodotus was not simply setting up an episodic chronology for Egypt like that of Lydia and Persia; he was establishing a linear sequence that would record the birth and growth of Egypt. He thus began the account on the theme of how the age of Egypt was learned, telling how Psammetichus had reared two children in absolute silence to find out what language they would speak. When they uttered what seemed to be a Phrygian word, he concluded that the Egyptians were not the oldest race on earth. The historian went on to explain that not only were the Egyptians younger than the Phrygians but their land itself had a beginning. He proved this in a variety of ways that depend on being able to measure the time in years between certain events in the past and Herodotus' own time. First of all we learn that all of lower Egypt during the time of Min, the first human king of Egypt, was covered with water. Herodotus claimed he learned this from the priests in Egypt, but he also sought out indirect proofs, noting that 900 years before his own voyage to Egypt, the land could be flooded by a rise in the river only half as great as that required in his own day. Proofs are found not only in the memories and records of Egyptian priests but in observations from his own experience. In his own day, shells on the mountains testified to their former exposure to water, and as sailors approached Egypt mud could be found on the bottom of the sea while the ship was still a day's sail from the land.

This change in the land is an important event for Herodotus. It brought into being not the Egyptian people, who had always existed, but the land on which they developed their culture.

> Then if there was once no country for them, it was but a useless thought that they were the oldest nation on earth, and they needed not to make that trial to see what language the children would first utter. I hold rather that the Egyptians did not come into being with the making of that which the Ionians call the Delta: they ever existed since men were first made; and as the land grew in extent many of them spread down over it, and many stayed behind. (2.15)

The time frame for measuring this event is fundamentally different from those used to measure the events in Asia and Greece. Those chronologies were tied to the spatial borders in which they obtained; here Herodotus created a time frame which measured the creation of space itself. This time frame is vague to be sure. Trying to decide how long it took to create the Delta, he estimated that 10,000 years would suffice to fill up the Arabian Gulf if the Nile were to change its course. When we compare this figure with the 900 years he thought it took to double the required level of water for flooding, we realize how unconcerned he was with precisely quantifiable terms. The vagueness of his numbers should not, however, blind us to the importance of Herodotus' insight. He drew the events of Egyptian history against the backdrop of a non-repeating linear process of natural change, a change which produced the very land which formed the setting of Egypt. Furthermore the time frame he established was suitable for another important linear change he discussed in Book Two: the appearance of Greek religion.

In establishing a linear time frame for how the Greeks learned about their religion, Herodotus displayed an interest in synchronizing important events from several different places, an interest conspicuously lacking from his account of those events dated episodically. He observed that Heracles was one of the gods 17,000 years before the reign of Amasis, that the temple of Heracles at Tyre was there at the beginning of the city, 2,300 years ago, and that his temple at Thrasos was built five generations before the temple of Heracles in Greece (2.43–44). This series of dates identified the antiquity of Heracles in three ways—relative to the Egyptian political chronology, to the appearance of the god in Greece, and to Herodotus' own day. The chronology is unlike any of the others in the work in that it forms in the reader's mind a single linear series beginning with the Egyptian land itself and continuing to the time of the author. As he proceeded with his account of Greek religion, Herodotus integrated other religious figures into this linear time. Reminding the reader of his earlier calculation of the years between Heracles and Amasis, he noted that Pan was older still, and Dionysius, the youngest, was considered by the Egyptians to be 15,000 years before Amasis. By contrast, the Greek Dionysius was about 1,600 years before Herodotus' own time; Heracles, about 900; and Pan about 800 (2.145). Thus the Greeks had learned about their gods later then the Egyptians.

The events Herodotus organized into a single temporal order were of a peculiar nature. They involved the religion which was so basic a part of Greek life, and they existed in linear time because they were learned in a process which could be tied to specific people and events. When Herodotus described the rites of Dionysius, he noted the similarity between the rites in Egypt and in Greece. The rites resembled one another, he thought, not because of the innate nature of the god, but because Melampus himself, who taught the Greeks about Dionysius, learned his art from the Egyptians (2.49).

Linear time was essential to the conclusion here. Herodotus was convinced that all the names of the gods came into Greece from Egypt. The foundation of his conviction was not a body of episodic anecdotes but his analysis of a series of events based on the linear chronology he established in this book. He concluded his discussion of Greek religion, "But whence each of the gods came into being, or whether they had all for ever existed, and what outward forms they had, the Greeks knew not till (so to say) a very little while ago [ou prōēn te kai khthes]; for I suppose that the time of Hesiod and Homer was not more than four hundred years before my own; and these are they who taught the Greeks of the descent of the gods" (2.53).

To accept the linear progression of religion Herodotus had to go beyond the evidence that had proved for him the growth of the land. In the land he could point to specific empirical data which could be seen in his own time and which suggested the existence of an orderly change from past conditions. In the case of the Egyptian origin of Greek religion, he found a linear progression of culture despite important differences between the two cultures in his own day. He faced the cultural differences between contemporary Egypt and Greece squarely, noting for instance that in Egypt the men do the weaving and the women do the marketing, a reversal from the Athenian practice (2.35). The affinity between the two peoples depended not on his own observation of present day similarities but on the reality of linear time, which allowed Herodotus to create a connection transmitted from the one religion to the other over a quantifiable period of years.

There is also an episodic chronology for Egypt in Book Two that lacks this linear quality. When he calculated the regnal years, Herodotus presented a picture of timeless duration in which the temporal relation among the events that occurred in each reign had no real significance.

Thus the whole sum [of years of each pharaoh's reign] is eleven thousand, three hundred and forty years; in all which time (they said) they had no king who was a god in human form, nor had there been any such thing either before or after those years among the rest of the kings of Egypt. Four times in this period (so they told me) the sun rose contrary to his wont; twice he rose where he now sets, and twice he set where now he rises; yet Egypt at these times underwent no change, neither in the produce of the river and the land, nor in the matter of sickness and death. (2.142)

This generational dating system clearly is not part of the linear one, where Herodotus had described a dramatic change in the land and river during the previous 900 years.

Herodotus, then, used two different dating systems which recorded events in two separate times. Most of the events in the work are included within a generational system which dated by successions of kings and rulers. The times measured by this system are episodic. The significance of events it measured cannot be seen in antecedent events but in thematic resonances which are not linear, that abruptly shift the reader in time and that give no precise origin for the narrative.

Beside these events in episodic time is another important group of events organized in linear time. To locate these events Herodotus used a dating system which synchronized two fixed points—Herodotus' own times and the reign of Amasis. These points do not have the precision that a modern historian would demand for a dating system, since Herodotus never clearly distinguished between the time he visited Egypt and the time he was writing, nor did he say what point in Amasis' reign he was dating from. But for the group of events he was describing such niceties were irrelevant. He successfully demonstrated to his readers that the culture of Egypt was much older than that of Greece and that this temporal relationship had an important significance, since Greece learned spiritual matters from Egypt. To reinforce this order of knowledge Herodotus joined to it an even longer order of natural events, portraying the land of Egypt as an event with a discernible beginning in time, which could be roughly measured quantitatively and which constituted a nonrepeating linear sequence.

Neither of these times is absolute; both derive their shape and measure from the subject matter. As Herman Fränkel has said of Herodotus' time, "It moves along as the events flow, stands still to effect a description, and turns around as he changes his attention from the son to the father."[9] In the case of episodic time the temporal status of any event cannot be identified until we know where it takes place. In the linear sequence the order depends upon the nature of the events it organizes. The sequence from Amasis to Herodotus' day is meaningful only in two ways, as a measure of the creation of the land and as an intellectual sequence whereby specific gods and religious practices are first learned by the Egyptians and then taught to the Greeks. The intellectual and geographical changes are significant, as is the order of events that produce them, but the absolute number of years is not. The order is entirely relative to the changes it describes; the chronology applies to a particular order and type of change and to no other.

Nor are these two times unconnected. They give different perspectives on the same theme, portraying both the disconnected and the developmental aspects of the conflict which Herodotus saw at the heart of the Persian wars. Neither could be eliminated from the narrative without robbing it of an important dimension. The existence of linear and episodic times enriches the story, allowing Herodotus both to make his themes immediately present to the reader and to give them depth and background. We seem to be present in Candaules' bedroom, witnessing the scene as intimately as Gyges himself and feeling his fear and embarrassment, just as we seem to listen to the very words of Xerxes as he tells of his dreams that lead him to ignore the doubts of his counselors and continue the expedition against the Greeks. Our involvement in these events would be lessened by the intervention of linear time and depth. For them Herodotus created a time of foreground where events need no antecedent conditions to give them meaning.

But such a time, standing alone, threatens to stultify interest in the past, to make the past a meaningless collection of anecdotes. When speaking of events in this episodic time Herodotus often made remarks that give us the feeling that such was the opinion of the Greeks. Promising to tell of all sorts of states, big and small, Herodotus explained, "For many states that were once great have now become small: and those that were great in my time were small formerly. Knowing therefore that human prosperity never continues in one stay, I will make mention alike of both kinds" (1.5). The presence of linear time redeems this stance and creates a genuinely historical point of view. We see the development of Greek religion and culture as part of a deep and pervasive temporal process, as old and ordered as the retreat of the waters that left shells on the Egyptian mountains and marked the times past that formed the very physical substance of Egypt.

THUCYDIDES AND THE LINEAR TIME OF WAR

Some might think that the paucity and inconsistency of source material lay behind these disparate chronologies, but Herodotus' times are only partly determined by his sources. The chronological material is based largely on lists of kings, which naturally tend to avoid synchronization with other reigns or countries. But the sources alone do not explain his practice. Despite the difficulties, he did synchronize events from several parts of the world in constructing the linear time sequence in the work, and despite the ease with which he could have synchronized the chronologies of Lydia and Mesopotamia, he failed to do so. Clearly the episodic and linear chronologies present separate facets of the story; the separate dating sequences he found in the sources were in his mind a reflection of reality.

Any suspicion that the distinctive traits of Greek historical writing can be traced to the type of sources they used can be allayed by considering Thucydides' history of the Peloponnesian War, for Thucydides not only created his own system for dating the events of the war but also told key parts of the story in episodic time. Since his subject, the long war between Athens and Sparta, concerned events within the Greek world, he had access to lists with which he could integrate the various theaters of the war. He used them to date the first hostilities but created his own system for the war itself, because he felt that the lists could not produce the kind of linear sequence he wanted.[10]

> But one must reckon according to the natural divisions of the year [kata tous khronous], not according to the catalogue of the names of officials in each place, be they archons or others who in consequence of some office mark the dates for past events, in the belief that this method is more to be trusted; for it is really inexact [ou gar akribes estin], since an event may have occurred in the beginning of their term of office, or in the middle, or at any other point as it happened.

The sequence Thucydides constructed for the war was clearly a linear one, where the order of events was essential to the outcome. It dated concrete events capable of simple location on a single time line. Because of this possibility for simple location, synchronization among the various dating systems of the Greek world was implicit in his method, and it is to that synchronization that Thucydides drew his readers' attention in the passage quoted at the beginning of this chapter (2.2).

The Peloponnesian War existed in linear time for Thucydides because the sequence itself had a meaning; he saw the important events in the war as consequences of antecedent events. When the plague broke out in Athens early in the war, Thucydides made clear its connection with the events that preceded it and those that came afterwards. He located the plague in time with reference to military events, saying that before the Lacedaemonian army had been many days in Attica the plague began to show itself among the Athenians (1.47). In describing the plague, he discussed the effect which the crowded conditions in Athens, produced by the Lacedaemonian invasion, had on its spread. After his account of the plague, he noted that it influenced the course of the war itself. Fear of it caused the invaders to leave Attica in haste, while sickness in the Athenian army impeded the siege of Potidaea. Finally he explained the Athenian loss of morale as a result of the combination of invasion and plague (1.57–59). His account of the plague depended for its meaning on the specific events which occurred immediately before and after it. For this reason he needed to organize the war by season and ignore the lists of officeholders, for the linear sequence was all-important for an understanding of the meaning of the story.

To portray the war as a coherent event in linear time Thucydides insisted that it was a single war whose component events existed in a single linear sequence. He maintained this position even when the events themselves did not cooperate. His main problem occurred in Book Five, where he had to discuss a truce of six years and ten months during which there were no direct hostilities between Athens and Sparta. When he resumed the narrative after this truce, Thucydides demonstrated his determination to continue the sequence.

> The history of these events [following the period of truce], also, has been written by the same Thucydides, an Athenian, in the chronological order of events, by summers and winters, up to the time when the Lacedaemonians and their allies put an end to the dominion of the Athenians and took the Long Walls and Piraeus. Up to that event the war lasted twenty-seven years in all; and if anyone shall not deem it proper to include the intervening truce in the war, he will not judge aright. For let him but look at the question in the light of the facts as they have been set forth and he will find that that can not fitly be judged a state of peace in which neither party restored or received all that had been agreed upon. And, apart from that, there were violations of the treaty on both sides. . . . So that, including the first ten-years' war, the suspicious truce succeeding that, and the war which followed the truce, one will find that, reckoning according to natural seasons, there were just so many years as I have stated, and some few days over. (5.26)

To reinforce the unity of the war and its location in time, Thucydides called to his support the aid of supernatural forces he usually scorned. He noted that oracles had predicted the war would last just twenty-seven years. In an earlier passage that identified the point at which the war began, he observed that it was accompanied by a series of unnatural disasters that made believable old stories previously thought fabulous, disasters which "fell upon them simultaneously [ama] with this war" (1.23). His willingness to use natural disasters to give emphasis to the war at its beginning and end stands in stark contrast to his usual skepticism about portents. When an earthquake struck Delos early in the war, he remarked dryly, "This was said and believed to be ominous of coming events, and indeed every other incident of the sort which chanced to occur was carefully looked into" (2.8). The same critical stance marks his response to the oracle which his contemporaries applied to the plague, which he said was an example of the general tendency of people to bend oracles to fit the circumstances (2.54).

But Thucydides saw in the Peloponnesian War more than a series of datable events, and he did not try to comprehend it entirely in his linear sequence. He saw it also as a conflict the parties to which manifest deep psychological differences. To understand the war, Thucydides realized that it was not enough to grasp the series of events; readers also had to see the psychological realities that lay behind them, realities which did not exist in the same linear time as the concrete events of war. He first brought these realities to the readers' attention in describing the events that led up to the war—the revolt of Epidamnus against Corcyra, and the Corcyraeans' subsequent appeal to Athens for help. In that section the differences in character between the Spartans and Athenians appear not as part of the sequence of events but as preconditions of the events, described in the orations which Thucydides used to explain them.

The Corinthians, bitter enemies of Athens, offered the most forceful description of the difference, when they gave an assessment of Athenian character to the Lacedaemonian assembly.

> For they are given to innovation and quick to form plans and to put their decisions into execution, whereas you [the Spartans] are disposed merely to keep what you have, to devise nothing new, and when you do take action, not to carry to completion even what is indispensable. Again they are bold beyond their strength, venturesome beyond their better judgment, and sanguine in the face of dangers. . . . In this way they toil, with hardships and dangers, all their life long; and least of all men they enjoy what they have because they are always seeking more, because they think their only holiday is to do their duty, and because they regard untroubled peace as a far greater calamity than laborious activity. Therefore if a man should sum up and say that they were born neither to have peace themselves nor to let other men have it, he would simply speak the truth. (1.70)

68

This character forms a vital connecting link among the events of the war. It determined Athenian victories early in the war, when by a bold move they isolated the Spartan army at Pylos, but it led them into calamity when, driven by the same boldness, they undertook the impossible mission to Sicily. Though Sicily was far too distant and too populous for the Athenians to hope to take and though the cautious general Nicias informed them clearly of the difficulties, Alcibiades persuaded the Athenians they must accept the challenge. He put their dilemma in terms that readers, by this point in the story familiar with Athenian character, could understand. Referring to a plea from their Sicilian allies and recalling earlier requests from similar allies back to the Corcyraeans' plea that began the war, he argued,

> On what reasonable plea, then, can we hold back ourselves, or make excuse to our allies there for refusing to aid them? We ought to assist them. . . . It was in this way that we acquired our empire . . . by coming zealously to the aid of those, whether barbarians or Hellenes who have at any time appealed to us. . . . For against a superior one does not merely defend oneself when he attacks, but even takes precaution that he shall not attack at all. It is not possible for us to exercise a careful stewardship of the limits we would set to our empire; but, since we are placed in this position, it is necessary to plot against some and not let go our hold upon others, because there is a danger of coming ourselves under the empire of others, should we not ourselves hold empire over other peoples. And you cannot regard a pacific policy in the same light as other states might, unless you will change your practices also to correspond with theirs. (1.70)

And when Athens suffered defeat at the hands of the Syracusans, Thucydides commented "of all the cities with which they had gone to war, these alone were at that time similar in character to their own" (7.55).

To describe the historical dimensions of this character, as well as the opposing character of the Spartans, Thucydides did not use the same linear dating he adopted for the war itself. Athenian character was an event, in that, like the hubris of the Asian kings, it had a temporal dimension; but also like that hubris, it was not entirely made up of primary events datable on a linear time line. To convey the Athenian character as something formed by a temporal process, Thucydides used an episodic time frame. After describing this character in the Corinthian speech quoted earlier, Thucydides focused on its historical roots in a section of Book One called the Pentekontaetia. This section takes up twenty chapters in Book One, about one-seventh of the book. In it he covered the fifty-year period (hence the name ascribed to the section) after the battle of Salamis in which Athens grew into an imperial power. This growth was of crucial importance to Thucydides, for he felt that the pretexts alleged by the combatants at the outbreak of the war masked its true cause. This he said was the "growth of the Athenians to greatness, which brought fear to the Lacedaemonians and forced them to war" (see 1.23). Though Thucydides used the verb "parekho" here to denote a causal relationship between the Athenian growth and the Spartan fear, he did not see the cause as a purely efficient one, to be explained entirely by the linear priority of the growth of Athenian power. He saw the growth and the fear as parallel developments out of differing character traits. The events recorded in the Pentekontaetia thus both explain the origins of the war and introduce the reader to the operation of Athenian and Spartan character.

Although Thucydides related many of the events in the period from Salamis to the outbreak of the war very precisely to one another, he did not use the beginning of the Peloponnesian War or any other event as a fixed date of reference. Instead he employed a series of temporal measures which give the reader an idea of how long particular wars and revolts lasted. Events dated thus have a relation to other specific events only when Thucydides wished to draw our attention to it. These indications of time are episodic and do not permit us to construct a single linear sequence for all the events of the period which he mentioned. Thucydides' use of an episodic dating system here has produced considerable dispute as to the actual chronology of the events being narrated.[11] The care he devoted to providing a clear linear sequence for the events of the war itself seems absent here, where he could certainly have provided a clearer and more precise dating than he did.

Not only did he leave the Pentekontaetia separate from the linear dating system he used for the war, but he placed his discussion of that period out of chronological order. Book One consists of four sections: (1) the early history of Greece down to the battle of Salamis; (2) the events immediately preceding the outbreak of hostilities between Athens and the Lacedaemonian Confederation; (3) the growth of the Athenian Empire between the battle of Salamis and the beginning of the Peloponnesian War; (4) the events immediately following the outbreak of war. Had these sections been placed in chronological order, the third would have come between sections one and two. Instead Thucydides placed his discussion of the Pentekontaetia between the breaking of the truce and the first hostilities of the war, near the end of Book One. He justified the digression with deceptive candor.

> I have made a digression to write of these matters for the reason that this period has been omitted by all my predecessors . . . and Hellanicus, the only one of these who has ever touched upon this period, has in his Attic History treated of it briefly and with inaccuracy as regards his chronology [tois khronois ouk akribōs]. And at the same time the narrative of these events serves to explain how the empire of Athens was established. (1.97)

Thucydides was thus concerned with the accuracy of his dates in the Pentekontaetia, just as in the war as a whole, but if we look at the organization of that section, it becomes clear that he saw the problem of temporal accuracy in two different ways—linear in the events of the war, and episodic in the Pentekontaetia.

The Pentekontaetia begins with a series of Athenian conquests, first of an island held by the Persians, then of land belonging to other Greek cities and finally of Naxos, a city prevented by Athens from breaking its membership in the league. Thucydides intended his reader to see the growth of Athenian aggressiveness, and he made that point explicit. "This [Naxos] was the first allied city to be enslaved in violation of the established rule, but afterwards the others also were enslaved as it happened in each case" (1.98).

Other revolts followed that of Naxos, as islands refused to pay their tribute or resisted military service. Before describing these revolts, Thucydides outlined the strategic situation of the dependent members of Athens' empire. Athens' allies had gradually chosen to offer monetary tribute rather than actual military service, allowing Athens to use the money to build its own navy. When the revolts began, the islands found themselves without military experience or equipment in opposition to a well-trained and well-provisioned Athenian navy. This process is a linear one, but no dating was necessary here, since the temporal progression from military readiness to dependence on Athens was clear from his previous list of Athenian actions.

The issue in the Pentekontaetia, however, is not so much the loss of independence by Athens' allies but the process by which Athenian growth produced fear in the minds of the Spartans. This fear had its roots in the revolts after Naxos, and Thucydides used an episodic time frame to describe the process. The first of these revolts was at Thasos, which for the first time he did not connect precisely to previous events, noting only that it occurred "after this [meta tauta]." He could have established a more explicit date for the revolt and in fact did so later in the work. The Athenian expedition to put down the revolt at Thasos went on to colonize Amphipolis, which Thucydides in Book Four said occurred thirty-two years after the first colonization and twenty-nine before his own day (4.102). Thus he could have placed the revolt at Thasos explicitly on a time line connected with his own day or he could have dated it with reference to the war itself, thirty-four years before it broke out. Here in Book One, however, he was not concerned to date the revolt at Thasos in linear time. Instead he wanted to show how it illustrated the growing Spartan fear of Athens which he saw as the root of the war. To this end he introduced at this point in the narrative a second revolt, an uprising of the helots against the Spartans at Ithome. The two revolts are connected by the fact that the revolt at Ithome prevented the Spartans from coming to help Thasos, which, unable to resist alone, had to accept Athenian terms after three years of war.

The scene now shifts to the Spartan troubles. Sparta asked for Athenian help against the helots, since the Athenians were so good at siege operations, but Athens' help gave rise to the first open quarrel between the two powers.

> For the Lacedaemonians, when they failed to take the place by storm, fearing the audacity and fickleness of the Athenians, whom they regarded, besides, as men of another race, thought that, if they remained, they might be persuaded by the rebels on Ithome to change sides; they therefore dismissed them, alone of the allies, without giving any indication of their suspicion, but merely saying that they had no further need of them. The Athenians, however, recognized that they were not being sent away on the more creditable ground, but because some suspicion had arisen; so because they felt indignant . . . they gave up the alliance. (1.102)

Thucydides did not place these two events, the revolt of Thasos and the siege of Ithome, in chronological order; instead he simply connected them with the indefinite particle "de" which conveys a general, unspecified relationship. They do not in fact form a sequence but are parts of a single episode, whose main outlines can be drawn in a single time frame as if they occurred simultaneously, since they produced a single result—the alienation of the Athenians and the Spartans. In both cases Thucydides told the duration of the revolts—three years for Thasos and ten for Ithome—but never gave dates that would allow precise temporal location within a single time frame. Many scholars today consider the ten-year span for the revolt at Ithome to be incorrect, since it is difficult to fit so long a revolt into other events known to be occurring during this period. Thucydides' primary focus here, however, was not with temporal location but with the duration of a group of episodes which first gave open expression to the fear that led to the war.[12]

Throughout the rest of the Pentekontaetia Thucydides brought in many complex relationships among the cities of Greece, all contributing to the general animosity between the Athenian Empire and the Lacedaemonian Confederation. Though he maintained some chronological order, he gave no dates and did not observe strict linear sequence. Words bespeaking a general relationship, like "kai de," predominate. Thucydides used them to present a thematic order undisturbed by considerations of rigid linear sequence.

The fifty-year period before the outbreak of the Peloponnesian War did not have the linear order of the war itself. The elements which produced the war occurred in a time frame more like those which Herodotus used to depict the fates of the kings of Asia. The realities had a temporal meaning, but they transcended strict linear sequence and could not be accurately portrayed by the sort of linear chronology Hellanicus used. Thucydides considered episodic time more accurate here. It portrayed better the actual complexity of the background by circumscribing events within groups and leaving open the question of progression from one group to another. The clarity with which we see Athenian domination and the Spartan fear depends on our seeing each story as a whole, and a linear sequence was inadequate for this, just as it was perfectly adapted to seeing the war itself.

The events of the Pentekontaetia happened long before Thucydides began to write of the war, but the distance in time cannot explain the fact that he dated the period episodically. Herodotus, after all, had used a linear sequence for the most distant events. Furthermore Thucydides himself used linear systems for the earliest events in his own works. To strengthen his claim that the Peloponnesian War was the greatest ever fought, he dated the events of early Greek history in a single sequence after the Trojan War (1.12). In addition he placed changes in shipbuilding and fighting techniques in a linear sequence going back from the end of the Peloponnesian War (1.13). In both cases he saw a linear order in which things were learned from an earlier period and in which, therefore, antecedent events had a direct influence on later ones. He saw the process of learning as a linear one, just as Herodotus had when he described the learning of Greek religion. In these cases a linear dating sequence, however vague and imprecise, was preferable to an episodic one.

Most modern observers would consider the process described in the Pentekontaetia to be a linear and developmental one, best presented in a linear series of dates. Certainly the growth of Athenian power seems to us to be a linear event. Yet the cause of the war went deeper than this single process in Thucydides' mind. Spartan caution and fear were as much a part of the cause as Athenian growth, and the fear and caution existed as historical events in their own right, not just as reactions to Athenian policy. Furthermore, the Athenian character that lay behind the growth was as fundamental a historical factor as the growth itself, and it did not exist as a linear sequence of events. By presenting the origins of these factors without a linear dating system, Thucydides brought them together as simultaneous and overlapping historical events, whose effect on the war depended on their mutual existence, not on the linear order of the events that created Athenian power and Sparta fear.

As foreign as they are to modern sensibilities, the dating systems adopted by Herodotus and Thucydides were not primitive strivings for an absolute time line. Had they been concerned with such a time line, they had available a variety of means to create one, including the lists of eponymous officeholders of the Greek city-states. They could not have achieved a degree of precision that would satisfy modern scholars by these lists, but they could have integrated events more explicitly than they did. They refused to use the lists because they wished to express changes in more than one time frame.

THE SHAPE OF RELATIVE TIME

The Greeks regarded events not as part of a single time but as grouped in several time frames that interacted only at specific points and in certain definite ways according to the purpose for which the historians were telling the story. Such a conception of time is relative in a mathematical sense. Though the relation between events within a series may be expressed quantitatively, the quantity does not indicate something absolute and independent of the relation. When the historians saw no interaction or relation, there was no need to synchronize events. They had no temporal relationship. A dating system presuming that any two events had only one possible quantitative temporal relationship and could thus be synchronized unequivocally would have been as inaccurate in the eyes of Herodotus and Thucydides as their relative dating is primitive in ours.

But there is a broader sense in which this time can be said to be relative. Croesus' recollection of Solon's warning, Sparta's distrust of Athens, and the growth of Athenian power are events measured in a time that is relative more in a psychological and literary sense than in a mathematical or scientific one. Herodotus and Thucydides saw as historically meaningful many temporal relations that could not be expressed in quantitative terms. In part the meaning they saw grew out of limitations in the sources available to them, but there are more positive factors involved. These relations also depended on imprecise and often unpredictable connections among events widely separated in quantitative time but immediately connected in real time. For such events episodic time was an appropriate vehicle of organization and presentation. The relation could be conveyed episodically, thematically, personally, or in any other framework suitable to the changes under consideration.

So entrenched is the idea that temporal order exists objectively, independent from the activity of the narrator, that we find it hard even to see the workings of another approach to time, one where the meaning determines the temporal order. That these historians made time depend on meaning can be seen in comparing two events told by Herodotus, the fall of the false Smerdis in Persia, and the battle of Mycale. In the first, he told events in episodic time, with an outcome that seems to defy our sense of temporal order; in the second he told them in linear time with results that defy our sense of natural causes for historical events. In both, the determining factor for Herodotus was the moral and social context of the events, not the linear order as an entity separate from meaning.

The story of the false Smerdis was part of the larger story of how Darius I took the throne of Persia, succeeding the family of Cyrus (3.67–69). It was for Herodotus an important example of metabole, that change in fortune that transferred power from one to another and that was a central concern for him throughout his work. The false Smerdis came to power in the first place through the insanity of Cambyses. When Cambyses went mad in Egypt, he sent his trusted lieutenant Prexaspes back to the capital to kill his brother Smerdis, convinced that Smerdis was plotting against him. Prexaspes carried out his mission, but Cambyses himself died before Smerdis' death became known. After news of Cambyses' death reached Susa, one of the Magians who had been administering the government took advantage of the confusion to seize the throne by claiming to be Smerdis. The false Smerdis resembled Cambyses' brother physically, and Herodotus (erroneously) thought they had the same name. The deception succeeded partly because Prexaspes, the man charged by Cambyses with the assassination, was understandably loathe to admit that he had killed the real Smerdis now that the mad king was dead. So the false Smerdis reigned for several months before he was overthrown by a conspiracy of Persian nobles who had discovered his deception.

Herodotus divided the story of his overthrow into two distinct processes, both occurring at Susa in the vicinity of the royal palace. The first centered around the conspirators, and the second involved Smerdis and his fellow Magians. He began telling the story of the conspirators first. Otanes, a Persian nobleman, suspected the fraud when he noticed that the new king had had no public audiences and had not met with any noble Persians. As it happened, Otanes' daughter was married to the late king Cambyses and was still part of the new king's harem. Though she had never seen the real Smerdis and was with the false Smerdis only in the dark of night, Otanes told her how to identify the false one by a distinctive physical trait. When his daughter confirmed Otanes' suspicions, he sought help at once in overthrowing the impostor and soon assembled a group of nobles, including Darius. Many of these had already had doubts about the new king themselves and were easily convinced of the necessity for action. To prevent discovery the conspirators decided to act immediately. They resolved to go the the palace at once and kill the pretender, confident that the guards would suspect nothing and would not hinder such a high-born group from entering.

Meanwhile the second process of overthrow had begun. Smerdis decided to take Prexaspes into his confidence, relying on the Persian's reluctance to confess his part in the murder and hoping that support from one so eminent in Persian society would solidify his hold on the throne. Partly by bribery and partly by arguments and threats, Smerdis persuaded Prexaspes to mount a tower next to the palace wall, call the people together, and tell them that the ruler inside was really the king's brother. Prexaspes climbed the tower as he had promised, but when all were assembled he exposed the entire fraud and admitted his murder of the real Smerdis. Exhorting the Persians to rise up and kill the man who had usurped the throne held by so many illustrious ancestors of the dead king, Prexaspes ended his speech by hurling himself to his death.

Meanwhile the conspirators were on their way to the palace. When they learned of Prexaspes death, they briefly considered desisting from their plan, but encouraged by an omen, they decided to persevere. The guards, as they had predicted, made no attempt to stop them, the Magians inside the palace were easily killed, and when they arrived in the throne room, Otanes and his associates successfully attacked and killed the false Smerdis.

Several aspects of this account raise questions in the minds of modern readers. The most difficult part to believe concerns the admission of the conspirators to the palace. Why did the guards let them in so easily? The death of Prexaspes had occurred some time before the entry, for the conspirators had acquired news of the event. They had even had time to consider the merits of postponing their action, a discussion which lasted long enough to raise tempers (3.76). News of Prexaspes' betrayal had certainly reached the false Smerdis, for at the time the conspirators entered the palace he was ensconced with his advisors to plan a stratagem for the situation the suicide had created. Given the existence of such a dramatic crisis, why was it so easy for the conspirators to enter the palace?

Many possible answers come to mind. Perhaps the Magians were in such a state of panic that they neglected even the elementary precaution of warning the guards not to admit anyone. Perhaps the guards themselves, hearing the news, turned on the false Smerdis and opened the gates to the conspirators in support of their plan. The conspirators might have told the guards they were entering the palace to help Smerdis in his crisis. The guards themselves could have been so isolated from the part of the palace wall where Prexaspes' death occurred that they were still unaware of it. All of these are possible, though Herodotus' account seems to preclude some of them. He noted, for instance, that the guards admitted Otanes and his party not only out of respect but also because they suspected nothing (3.77).

In any case it would be easy enough to resolve the improbabilities of the event. Even if Herodotus found no answer in his sources, he could speculate on what might have made it possible. Instead he simply described the entry as happening "under heaven's guidance [theiē pompē khreōmenous]" (3.77). The easiest solution would have been to present the events as if they had no temporal interval separating them, occurring so close to one another as to be simultaneous in the Newtonian sense of the word. Herodotus' narrative precludes that, since he specifically noted that both parties knew of the suicide and each took steps to meet the new situation. The entry of the conspirators took place after the death of Prexaspes, although the death itself precluded the possibility that the entry could have occurred the way Herodotus described it.

Since the two stories do not fit well into a single linear time, some modern scholars have tended to solve the problem by ignoring the story of Prexaspes, narrating the conspiracy as a single process which went on without any real response from within the palace. Others simply note that there are two versions of the story.[13] Neither of these modern solutions is wholly satisfactory, for Herodotus interwove the stories in such a way as to suggest that in his mind at least they formed a single episode. He may have found them in different sources, but he brought them together into a single narrative. Furthermore his language in connecting them suggests that they existed in a network of causal and temporal factors that cannot be entirely specified. When he first left off the story of Otanes to begin that of Prexaspes, he said, "While they were thus planning, matters fell as I will show" (3.74). The word Herodotus used to express the relationship is "suntuchia," which has connotations of accident and conjuncture. It left the precise relation between the two processes unspecified.

Many themes are blended in the story of Smerdis' downfall, evoking in the reader's memory echoes of Herodotus' account of the early kings and laying the groundwork for the confrontation between the Arachmenids and the Greeks which would dominate the book. We see the remoteness of the eastern kings from their subjects, the arrogance and intrigue of the court, the boldness and decisiveness of the first Arachmenid king. Both stories of Smerdis' downfall contribute to these themes; each paints a distinctive picture of the episode. Herodotus was willing to sacrifice neither story to the demands of a linear sequence and had available a sense of time which allowed him to keep both.

In Herodotus' account of the battle of Mycale, by contrast, he depended on a precise linear sequence of events to explain the outcome. The fall of Smerdis involved parallel episodes which modern readers feel the need to make into a single linear story, but the battle of Mycale depends on a linear sequence which we would consider impossible. In that battle, which took place in Ionia between Greek and Persian forces, the Persian defeat occurred a few hours after the Persian army in Greece was defeated at Plataea. Herodotus took great pains to establish the order of those two battles, noting that the battle at Plataea happened early in the day, while the Persians were beaten at Mycale in the afternoon. Moreover, he assured us, Greeks looking into the matter after the battles established that both had occurred on the same day of the month (9.101).

The importance he ascribed to the temporal order of the two battles is clear. Shortly before the troops engaged at Mycale a rumor of the victory at Plataea circulated in the Greek army and played an important role in defeating the Persians. To highlight the effect of that psychological advantage at Mycale, Herodotus in telling the story of Plataea stressed the Greek strategic advantages rather than their moral qualities. He portrayed as the decisive factor at Plataea the superiority of Greek arms. He felt the Persians were the equal of the Greeks in courage and physical strength, but they fell because they were struggling without proper armor against heavy infantry (9.62).

At Mycale the material advantages that had been such a crucial factor at Plataea had little to do with the outcome. From the very start Herodotus portrayed the battle as a psychological one. Before combat the Greek commander, sailing up to the walls of Mycale, urged the Ionians to desert from the Persians in the coming battle. For fear of such a desertion the Persians detached Ionian troops from their forces, thus weakening their army. Against the background of these psychological issues, Herodotus described the effect of the rumor about the victory at Plataea. "Now before this rumor came they had been faint-hearted, fearing less for themselves than for the Greeks with Pausanias; . . . but when the report sped among them they grew stronger and swifter in their onset. . . . When the Athenians and their neighbors in the line passed the word [of the victory at Plataea] and went more zealously to work . . . immediately the face of the fight was changed" (9.101–2).

To a modern reader the key event here is the rumor of victory, not the coincidence of the two battles. Plataea and Mycale were too far separated in space for news to be transmitted in the few hours between the two battles, so their actual temporal relationship was incidental and did not enter into the final result. But Herodotus considered the linear time of the two battles to be an essential part of the story. To explain the rumor he had to invoke a supernatural power,[14] but he also felt the need to provide a visible context, for the event existed in space and time. For example he showed his readers the spatial context within which this divine power worked. A herald's wand was found on the shore—showing that someone had brought news—and both battles took place in the precincts of a temple of Demeter, who had been the instrument of the communication.

Thus Herodotus constructed a linear temporal sequence to explain the victory at Mycale, locating the events on a spatial/temporal continuum. He saw the battle as one continuous series of events, with each event existing in clear temporal relationship to the others and only one at a time occupying the "present"; the revolt at Susa was a multiplicity of simultaneous occurrences whose events had no definite linear order. The spatial context of the linear time at Mycale included the whole eastern Mediterranean in one local time; in Susa the conspirators and the Magians occupied separate spaces, each with its own local time.

Neither of these times was absolute; each depended on the context and meaning of events. The order of priority in which he made temporal judgments for the battle of Mycale was the same as with the fall of the false Smerdis, though the result was different. In each case the meaning of the events dictated their temporal relations. The moral significance of Mycale demanded a clear linear order in a well-defined spatial field; that of Smerdis, an episodic sequence in which the linear aspects of time and space were absent. Consciousness of the antecedent victory at Plataea affected the result at Mycale; consciousness of Prexaspes' betrayal did not affect the result at Susa.

Two sorts of processes seemed particularly suited to linear time, those involving physical change, such as the growth of the Nile and the move-ment of the herald's wand at Mycale, and those involving the transmission of knowledge, such as the Greeks' acquisition of the gods from Egypt or the development of naval technique. Other processes were linear or not de-pending on their own shape. Political and military changes did not have to exist in linear time, though Thucydides created a single linear sequence for the war, which was his principal subject. Individuals did not necessarily exist in linear time. Their character was subject to dramatic changes, but these changes were not part of a sequential process in which antecedents led to consequences.

Croesus, for instance, after a reign marked by foolish interpretation of oracles leading to his defeat by Cyrus and the loss of his kingdom, became a trusted and wise advisor of the Persian kings, a man who looked deeply into the consequences of acts and saw in the affairs of his masters the pit-falls and opportunities he missed during his own reign. The change is an alteration of status, not an individual development. Croesus did not learn from a series of experiences; he changed abruptly and dramatically as he lay on the pyre Cyrus had already lit. As with the false Smerdis, Herodotus presented the change in ways that cannot be easily incorporated into a linear series. In one sense Croesus acquired his wisdom from understand-ing the general pattern of events that had led to his fall. He learned this pattern when, after his downfall, he sent the chains Cyrus had bound him with to the Delphic Oracle to complain of the poor advice he had been given. The oracle replied by pointing out all the patterns that converged to rob Croesus of his throne: the vengeance of the gods for the ancient crime of Gyges, the postponement of the fall of Sardis for three years so that Croesus could enjoy his kingdom as long as possible, and Croesus' own fault in not asking the oracle which kingdom would fall if he went to war with Cyrus. The explanation gave Croesus wisdom, for when the Lydians returned to Sardis with the Priestess' answer and reported it to Croesus, "he confessed that the sin was not the god's but his own" (1.91).

This insight into the series of events before his fall did not actually produce Croesus' newfound wisdom, for immediately after he was freed from the pyre and before he had had occasion to reproach the oracle, Croesus gave Cyrus excellent advice about how to hold Lydia and prevent a revolt among the Persians (1.88–90). The change in Croesus' character is not a process in linear time. The linear change is entirely a change in status which has nothing to do with Croesus the individual. Croesus the king took on reality for Herodotus as an illustration of metabole, a phenomenon that was particularly striking among kings because of their exalted wealth and power; Cyrus the advisor took on reality as the wise private citizen whose analysis of circumstances was so acute that it robbed the kings of excuses as they fell. He played the same role in the reigns of Cyrus and Cambyses that the Delphic Oracle and Solon had played in his own reign.

Even when events existed in linear time, they could be part of two different and incompatible linear sequences. During the Sicilian campaign the Corinthians defeated the Athenians in a naval battle at Naupactus by reinforcing their prows and ramming the Athenian ships head-on (Thucydides, 7.43).[15] Later the Syracusans, faced with an Athenian fleet at Syracuse, "had prepared the fleet generally in such a way as, after the experience of the former sea-fight, seemed likely to offer some advantage, and in particular had shortened the prows of the ships and had made them stouter" (7.36). Thucydides thus created a linear sequence in which advances in naval technology are learned from previous experience, a sequence which he began in Book One at the very beginning of Greek history. But in the linear sequence of military events that constitute the Peloponnesian War, this learned experience did not have the same shape. No Syracusans were present at the battle of Naupactus. The "experience of the former sea-fight" did not exist for them. It existed only for Thucydides, who as a historian saw the connection between the two battles and made it part of the pre-existent linear time of acquired naval technique. In his actual description of the battle, he made clear that the prows were of less importance than the action of the Syracusan javelin throwers, who took advantage of the maneuverability of the Syracusan light ships (7.40). Thus the naval battle between Syracuse and Athens existed in two linear times, one of immediate military strategy in the war itself and another of great duration in which naval technique improves as part of an essential process in time.

The subject matter determined the shape of Greek time. A particular war, polity, people, or religion created its own temporal boundary; a particular lesson, its own time frame. Neither of the historians whose works dominate the landscape of Greek historical writing conceived of events transcending their immediate experience of Greek culture and polity. Thucydides thought of the war as an event larger than any particular Greek state, but he developed a chronology based directly on the war itself to express its linear temporal dimension. The concept of a time line indefinitely extendable and applicable to several polities and several wars is not

81

present in Greek historical writing. Such a concept waited upon the integration of political and cultural life that accompanied the establishment of the Hellenistic Oecumene. Even that integration occurred against the background of the sense of time discussed here. Hellenistic historians, like their Hellenic predecessors, incorporated a variety of linear and episodic times into their narratives to convey the full depth and richness of their subject matter. An analysis of Hellenistic historians will reveal a similar commitment to relative time.

4

The Time of The Oecumene

The date from which I propose to begin is the 140th Olympiad. . . .
Previously the doings of the world had been, so to say, dispersed, as
they were held together by no unity of initiative, results, or locality;
but ever since this date history has been an organic whole, and the
affairs of Italy and Africa have been interlinked with those of Greece
and Asia, all leading up to one end. And this is my reason for beginning where I do.

Polybius, *The Histories*, 1.3

POLYBIUS AND THE OLYMPIAD YEAR

Written by an admirer of Thucydides two centuries and a half after his
death, this passage adopts a dating system remarkably different from those
Herodotus and Thucydides had used. Instead of the generations of kings
that marked time for Herodotus or the cumbersome and complex techniques Thucydides used to locate the beginning of the war, Polybius employed a single date, the Olympiad that marked the quadrennial celebration of the Olympic games. This chronology was not tied to any particular
political event in his narrative but stood apart from all of them, an independent measure of time.

Polybius did not invent the Olympiad dating system nor did he regard it
as the distinctive feature of his own work. "For what gives my work its
peculiar quality, and what is most remarkable in the present age is this.
Fortune having guided almost all the affairs of the world in one direction
and having forced them to incline towards one and the same end, a histo-

rian should bring before his readers under one synoptical view the operations by which she has accomplished her general purpose. Indeed it was
this chiefly that invited and encouraged me to undertake my task."[1] The
single end Polybius referred to and the event he sought to record was the
Roman conquest of the Mediterranean, an achievement he considered un-
precedented and of obvious interest and import. "Who is so worthless or
indolent as not to wish to know by what means and under what system of
polity the Romans in less than fifty-three years have succeeded in subject-
ing nearly the whole inhabited world to their sole government—a thing
unique in history [o proteron oukh eurisketai gegonos]?" (1.1)

Polybius saw in the Roman conquest not simply a major event worth
remembering but one which had permanently changed the course of history. He felt that the Roman conquest could be understood only by looking at human history as a whole and trying to find the unity which underlay it. Without grasping this unity he felt he could not explain how Rome
overcame so many obstacles to create a political unit out of the competing
states which had hitherto dominated the Mediterranean. Where previous
historians had focused on single wars or nations, Polybius thought in terms
of the whole and urged his reader to contemplate the synoptical view that
he presented in his narrative.

A single time line and a sense of historical unity transcending any particular event are vital parts of our own consciousness of time, but the very
fact that they are so indispensable to us makes it hard to understand how
Polybius used them. Modern readers tend to see the two elements, the
single time line and the sense of thematic unity, as interdependent parts of
a unified time, but Polybius did not. He used the single time line not as a
universal index of all events but as a limited framework whose function
was to measure the newly perceived unity of those historical events surrounding the rise of Rome. He felt that the new chronology offered specific advantages over other systems, and he organized his story around it
only where those advantages were present. In other cases he used different
principles of organization, principles which did not assume a single chronological order.

In telling the story of Rome's rise to power Polybius portrayed an event
which bore upon his own life. His family had been part of the leadership
of the Achaean League, the alliance of Greek states Rome defeated in
168 B.C. to assume control of the Greek world. After this defeat, the invaders captured Polybius himself and took him back to Rome as a slave.
Only after a long captivity did he return to his native Greece as an admin-
istrator for his captors. Out of his own experience of Roman power arose
his curiosity about its causes and antecedents.

But the unification was deeper and broader than even Polybius knew. Rome's advance into the eastern Mediterranean was only the political capstone of a cultural and economic process that had started many years before. In the century and a half before the defeat of the Achaean League, the conquests of Alexander the Great, together with the growth of commerce in the Mediterranean basin, created a common culture based on the Koinos, the Greek dialect spoken throughout the region. This Hellenistic Oecumene produced a sense of unity that lay behind Polybius' synthetic vision of history and inspired his narrative as deeply as his own participation in the resistance to Rome did. He both recorded and reflected the unifying trends of his time.

Partly because of the Hellenistic Oecumene, Polybius could see a broader chronological scope and from this vantage point could invoke more general principles of historical process. He felt that there was a pattern of recurrence in human affairs, a pattern more well-defined than the simple rise and fall that Herodotus talked of. In his most famous statement of this pattern Polybius explained that there was a cycle of governments, with good constitutions degenerating into bad ones. The three good ones, monarchy, aristocracy, and democracy, were each followed by an evil counterpart, tyranny, oligarchy, and mob rule. The cycle began with monarchy and ended with mob rule, which in turn produced another monarch to bring order back to civil affairs (6.4–9).

If Polybius' model for history were as simple as this, then a single time line would have served him very well to express the essential qualities of history, for the absolute temporal relations between two events would identify what changes of government were transpiring and indicate the direction in which they were going. But Polybius introduced many complexities which could not be encompassed in a single time frame. Most important, Rome itself had escaped this cycle through a mixture of constitutional forms. By combining elements of monarchy, aristocracy, and democracy, Rome had checked the fatal weaknesses in each form and given itself a degree of permanence that in part explained its conquests.

In addition to this major exception to the cycle, Polybius introduced into his narrative several other senses of historical recurrence. In an insightful study of Polybius, G. W. Trompf has found that his interpretation of events reflected several different patterns adopted from ideas of recurrence in the classical world. He incorporated biological notions of growth and decay in his consideration of previous empires. In the detailed narrative of Rome's wars he presented a pattern of reciprocal alternation around a balanced mean. And he introduced concepts of retribution to encompass the all-important moral dimension of history, describing particular achievements as either punishments or rewards for particular patterns of behavior.[2] Given the complex temporal dimensions he saw in the rise of Rome and the unification of the world, none of the systems used by the classical Greek historians was suitable for dating this development.

These conceptual issues apart, the scope alone of the events he wished to date made traditional chronologies inadequate. The lists of eponymous officeholders of the individual Greek city-states suffered from all the drawbacks Herodotus and Thucydides found in them. In addition, the rise of Rome unified several different polities, each with its own dating system. Roman lists of officeholders were available but were unfamiliar to the Greek audience for which Polybius intended his work. Thucydides' practice of dating from the beginning of the Peloponnesian War suggested using the Roman conquest as a starting point, but that would blur the significance of the antecedent events. More importantly, it would lose sight of the uniqueness of Rome. Polybius wanted to show that this was not just another war, however important and prolonged, but a cataclysmic change in human relations, permanently altering the interconnection of events in the world. He needed a chronology which was tied to no specific previous event and which would include the antecedent conditions of the Roman conquest.

The Olympiad was tied to no particular political unit, yet by using it to synchronize events in the different parts of the Mediterranean Polybius could show the importance of past events in creating the unified world. He acknowledged his debt to previous historians, observing that his predecessor Timaeus "compares the dates of the ephors with those of the kings in Lacedaemon from the earliest times, and the lists of Athenian archons and priestesses of Hera at Argos with those of the victors at Olympia, and who convicts the cities of inaccuracy in these records, there being a difference of three months" (12.11). Timaeus' history has not survived into modern times, and Polybius' description makes it hard to tell whether his predecessor actually drew up a comprehensive dating system based on the Olympic lists or whether he simply used them haphazardly. In any case the Olympiad had assumed the dimensions of a chronological reference point for some parochial dates by the time Polybius wrote.

Even before Timaeus, the Olympiad occurred in the writings of the classical Greek historians, but as an event in its own right rather than a pure date. When Thucydides dated a Lacedaemonian conference by referring to the Olympiad, he paused before describing the conference to note that Dorieus of Rhodes won his second victory in those games (3.8). The other time he mentioned them, when violations of the Olympic truce brought the Spartans into new hostilities, the games were part of the story (5.49).

Xenophon, who continued Thucydides' narrative, also cited the games occasionally, but, like his predecessor, never allowed them to stand alone as
the sole chronological reference or to constitute an abstract date (*Hellenica*,
1.21, or 2.31).

It was not in fact a historian who first used the Olympiad as an abstract
date. Eratosthenes, a third-century Alexandrian mathematician, developed a dating system based on the Olympiad in two works, neither of
which has survived. In the *Olympionikai*, he compiled lists of the games
and contestants to which he added other historical events from the beginning of the games down to his own time. In a more theoretical work, the
Khronographiai, he numbered for the first time the single years within the
four-year period of the Olympiad and created a chronological system based
on the resulting number series. From Eratosthenes' *Chronography* the dating systems of the West have sprung.[3]

Although Eratosthenes' *Chronography* itself is not available, what we
know about it and about his other interests strongly suggests that he conceived the Olympiad as an abstract and absolute dating system, independent from any particular series of events, and as a means of treating time
arithmetically. His other contributions all testify to his interest in quantifying and abstracting experience. He invented the sieve of Eratosthenes,
a procedure for locating all the prime numbers in any numerical series; he
also devised the meridian as a means of locating any place on earth on a
conceptual grid. The little we know about his chronology suggests that he
looked at the Olympiad as an abstract measure of time just as he conceived
the meridian as a measure of space. He is credited with being the first to fix
the fall of Troy as the earliest secure date in Greek history. Since the
Olympic lists began over 500 years after the date he assigned to the fall of
Troy, he certainly conceived of his time line as an abstract one, capable of
measuring events forwards and backwards with no fixed limit, just as his
sieve could find the prime numbers in any conceivable series.

The Olympiad was conceptually suited for creating an abstract time
line, but it still presented practical difficulties to a historian who wished to
use it for a particular series of events. It could never be an absolute indication of temporal location, since the games which determined its beginning
were calculated on a lunar calendar and not held on a fixed date in the
solar year. Even today historians cannot date the games more precisely
than as occurring at a full moon in midsummer.

Even after accepting these limitations of computation, a historian
would encounter further technical difficulties. From a military point of
view, the fact that the Olympiad year began in mid-summer posed problems. Since military campaigns opened in the spring, Polybius would have

had to introduce breaks in them in order to maintain organization by Olympic year. No narrative could clearly explain the military aspects of the Roman conquest while interrupting campaigns to preserve the integrity of an arbitrary temporal boundary, and Polybius made no attempt to do so. His usual practice was to close the year with the end of the military season. The length of the year thus depended on that of its military operations, not on the phase of the moon which dictated the celebration of the Olympic games.

There were also political problems that made it hard to use the Olympiad. Since Polybius recorded decisions and actions of civil bodies which did not themselves use the Olympiad, he found it hard to integrate the public records with the Olympic lists. These problems are clearest in the first part of his history, where he had to describe events which were identified in his sources only by local dates with no oecumenical or universal reference. Some scholars have suggested that he made a list integrating the dates of Carthage, Spain, Rome, Greece, and the Near East, but if he did so he made no use of his list in the narrative. In the first two books he mentioned the Olympic year only six times. Instead of following a uniform dating system he adopted the various systems found in his sources. Sometimes he used the year of the war he was describing;[4] other times he referred to the consular elections,[5] although he most often noted the actions of the new consuls without saying when they were elected, thus subordinating the date to the specific acts of the consuls (1.25,26). In the eastern theater he tended to use the general years of the Achaean League.

None of these problems posed an insuperable barrier to using the Olympiad as an absolute dating system. All such systems are arbitrary, at least in part, and by adopting a few conventions about the beginning of the year and its relation to various systems contained in the sources, Polybius could have implemented Eratosthenes' theoretical construct. Some scholars have indeed maintained that such was Polybius' goal and that he treated the Olympiad as an absolute time line and conceived of chronology as an arithmetical problem.[6] Others disagree with this point of view, maintaining that Polybius used a variety of chronological systems in addition to the Olympiad and never regarded that system as an all-inclusive one.[7]

Perhaps Polybius did understand and appreciate the absolute, arithmetic quality of Eratosthenes' system. Whether he did or not, he certainly did not look on that chronology as a thing in itself. He regarded it only as a useful tool to express and clarify the unique event he wished to portray—the Roman unification of world affairs. Only after the unification had occurred was it important to bring all the events into a single chronology.

Describing Philip of Macedon's intervention into Greek affairs, Polybius explained his procedure:

> This took place at the same time that Hannibal, after subduing all Iberia south of the Ebro, began his attack on Saguntum. Now had there been any connexion at the outset between Hannibal's enterprise and the affairs of Greece, it is evident that I should have included the latter in the previous Book, and, following the chronology [akolouthountas tois kairois] placed my narrative of them side by side in alternate sections with that of the affairs of Spain. But the fact being that the circumstances of Italy, Greece, and Asia were such that the beginnings of these wars were particular to each country, while their ends were common to all, I thought it proper to give a separate account of them, until reaching the date when these conflicts came into connexion with each other and began to tend towards one end—both the narratives of the beginnings of each war being thus made more lucid, and a conspicuous place being given to that subsequent interconnexion of all three, which I mentioned at the outset, indicating how, when, and for what reason it came about—and, then upon reaching this point to comprise all three wars in a single narrative. The interconnexion I speak of took place towards the end of the Social war in the third year of the 140th Olympiad. After this date therefore I shall give a general history of events in chronological order. (4.28)

Polybius clearly subordinated the single linear order created by the Olympiad to the story of Roman conquest and unification. For Polybius time did not measure the Oecumene; time itself, as a single measurable unit, was created by the Oecumene. Before the Roman conquest there were times, each measured in its own way and each possessing its own order and significance. After the conquest there was a single time, needing a new standard of measurement. The time line did not give meaning to the events; they gave meaning to the time line.

When he grouped the events after the 140th Olympiad into a single sequence, Polybius imposed an order the participants in those events had not seen and that his readers did not find obvious. The narrative thus appeared unnatural to many of his readers.

> I am not unaware that some will find fault with this work on the ground that my narrative of events is imperfect and disconnected. For example, after undertaking to give an account of the

siege of Carthage I leave that in suspense and interrupting myself pass to the affairs of Greece, and next to those of Macedonia, Syria, and other countries, while students desire continuous narrative [to suneckes] and long to learn the issue of the matter I first set my hand to. (38.5)

In defense of his practice Polybius pointed out that he left his readers the liberty to construct in their own minds a continuous narrative. He provided instead an understanding of the whole which could come only from treating the various affairs of the world in a uniform fashion.

But I myself, keeping distinct all the most important parts of the world and the events that took place in each, and adhering always to a uniform conception of how each matter should be treated, and again definitely relating under each year the contemporary events that then took place, leave obviously full liberty to students to carry back their minds to the continuous narrative and the several points at which I interrupted it, so that those who wish to learn may find none of the matters I have mentioned imperfect and deficient [ateles mēd'ellipes]. (38.6)

Most authors added digressions to give variety and interest; Polybius broke up his narrative only to promote a deeper understanding of the whole.

He saw the organization into a linear series of years as a means of bringing out the significance of his main theme, which was only one part of his story. This functional approach to the Olympiad system explains his occasional abandonment of chronological order, even in the later sections of his work, long after the time when he felt the unification of human affairs had taken place. During the 155th Olympiad the city of Oropus underwent a complex series of vicissitudes as the result of a quarrel with Athens. Rome intervened first on one side, then on the other; the inhabitants abandoned and then resettled the city. Polybius explained why he told the story out of order:

I will give a succinct account of the whole of this matter, partly recurring to the past and partly anticipating the future, so that, the separate details of it being by no means striking, I may not by relating them under different dates produce a narrative both obscure and insignificant [eutelē kai asaphē]. For when the whole seems scarcely worth close attention, what chance is there of any student really making it an object of study when it is told disjointedly under different dates? (32.11)

Polybius wrote with a clear purpose in mind. Like Thucydides before him, whom he regarded as a model, he wanted his story to bring practical wisdom, and he altered his narrative when it threatened to run counter to his goal. Both historians looked for a truth that transcended the events. For both, this truth depended partly on placing the events in a clear temporal sequence so that the antecedent conditions could be understood. Polybius departed from Thucydides because he was convinced that a fundamentally new event with permanent consequences required a temporal sequence not provided by any previous chronology. Thucydides, after locating the beginning of the Peloponnesian War by reference to existing dating systems, felt comfortable narrating by an internal dating sequence that would have no applicability beyond the war itself. Polybius, secure in the belief that the unity of the world would persevere, created a new dating system to express that unity. For both historians, however, the dating schemes had no independent existence. The chronologies served the deeper purpose of the narratives and derived their value from the clarity with which they expressed the wisdom which could be wrung from the events.

When insight into the whole necessitated a narrative that ignored the chronological order, Polybius did not flinch but pursued his ultimate goal at the expense of the linear series. After describing the battle of Cannae, where Hannibal annihilated the Roman army, he concluded Book Three with a chapter explaining its aftermath. There he neglected some events that immediately followed the battle and included others that occurred some time later. According to his narrative, the Tarentines revolted against Rome immediately after the battle, and the Celts massacred the Roman army in the north a few days later. In fact the Celtic victory did not occur until the following year, while the Tarentines did not revolt until three years after Cannae. Not only does the narrative place these disasters right after the battle, but Polybius ignored Scipio's victories over the Carthaginians in Spain that happened about the same time as Cannae.

The reason for this abandonment of chronological order is clear. The 140th Olympiad, when Cannae occurred, marked the starting point of his story—the beginnings of the Roman conquest and the growth of Roman power. In order to make the story a continuous linear process to fit his dating scheme, Polybius wanted to emphasize the unrelieved disaster from which the Romans had to recover. Concluding his account of the events after Cannae, he said, "For though the Romans were now incontestably beaten and their military reputation shattered, yet by the peculiar virtues of their constitution and by wise counsel they not only recovered their supremacy in Italy and afterwards defeated the Carthaginians, but in a few years made themselves masters of the whole world" (3.118).

Three separate concerns affected Polybius' sense of time. First, he agreed with Thucydides that a narrative organized year by year could best express wisdom that the past had to offer. He used the Olympiad year to synchronize events from different parts of the Mediterranean and show his readers that common forces and principles were at work in the complex workings of the Roman conquest. Second, he located these events on a time line unique in its scope, a time line devised by a mathematician for universal applicability. The Olympiad made possible the synchronization not only of events actually related in the mind of the historian, but of all events whether their connections were immediately obvious or not.

Third, Polybius' most important concern was the new unity in human affairs arising out of the unique event of Roman conquest and having a permanent and universal effect on his world. He wrote primarily to explain this event, intent on losing neither its historical uniqueness nor its conceptual universality. Because this unity had historical roots, Polybius chose a chronology that transcended its beginning; because the unity included several states, he used a chronology which easily synchronized events in different parts of the Mediterranean. These factors determined his use of the Olympiad to create a linear series of years as a framework for his narrative.

Annalistic form, a universal time line, and a sense of unity in historical events are separate ingredients in his work, though Polybius used them together to great effect. In combination they tended to reduce the episodic quality of the *Histories* and to stress the linear dimensions of the work. He was methodically concerned with presenting the antecedents of a single event. Although he saw several lines converging on this event, and even treated such material as the aftermath of Cannae episodically, he did fashion most of his material in linear sequences. He did not, however, blend these sequences into an absolute time line but used them only when they served his overall purpose of finding wisdom in a study of the past.

Since these three components of Polybius' time are conceptually distinct, it should come as no surprise that they underwent separate development among Hellenistic historians. All three proved useful, but during the half-millennium between Polybius and the decline of Roman political authority in the West, they served different functions and were seldom viewed as ideas that could be integrated into a single historical narrative.

Polybius' successors among Hellenistic historians, without his overwhelming commitment to a synoptical view of the course of events, moved in different directions. Some, including Livy and Tacitus, focused on the history of Rome, adopting an annalistic form because of its advantages in synchronizing events across the far-flung reaches of Roman conquest. These historians had little interest in creating an abstract linear series and marked the passage of years by using lists of political officials. Others, like Castor of Rhodes and Eusebius, developed new chronological systems. As broader and more complex vistas of time and space opened to the inhabitants of the Oecumene, the value of a single integrated chronology became clearer and clearer. Several chronologists sought to create a linear series of events that would blend in a single accessible form all of human history. Even these scholars, however, avoided a single dating system, and when they wrote their own historical narratives made little use of the chronological tables they had made. Finally a third group, including Pompeius Trogus and P. Annius Florus, focused on the conceptual unity of the world. Polybius' vision of a synoptical view in which all the events of human history could be recorded was not without its attractions in an age increasingly tied together by cultural as well as political and economic bonds. These historians looked for underlying trends and connections that would show an overall meaning to human history behind the particular events. In their quest, however, they avoided annual dates, describing the temporal relationships among the events in vague and entirely relativistic terms.

All these historians and chronologists made important contributions to our sense of time, but none created a single, absolute time line. Their efforts remained focused on the task of recording and understanding the Oecumene. To accomplish that task they took advantage of the flexibility afforded by relative measures of time, bringing out the distinctive temporal aspects of any particular event by viewing it in the context of a time appropriate to it.

ANNALISTIC FORM

Since annals were among the first historical records, the annalistic form often suffers condescending treatment as a primitive type of narrative. Yet some of the finest and most sophisticated historians in the Western tradition have organized their material annalistically, narrating events in the year in which they occurred rather than following themes over long periods of time. Livy, under the patronage of Augustus in the early empire, told the story of the Roman Republic from its beginnings to his own time. This monumental work in 142 books, of which 35 survive, presents the events year by year in a series of 744 years. Tacitus, whom many consider the greatest of the Roman historians, after a career in which he wrote monographs and a history of the later emperors organized by reign, wrote as his final work and his greatest achievement a history of the early empire organized as an annal.

In writing their histories as annals, Livy and Tacitus adopted a practice that militated against the construction of a single linear series. Instead they tended to present the events in relative time, looking at them episodically within the year rather than in the linear order of a number of years. They did not use an abstract date like the Olympiad to mark the boundaries between years but indicated the year by the election of the consuls. Lacking an absolute reference point, the individual years had no necessary or quantitative connection with one another.

It is easy to see why Roman historians rejected the Olympiad, with its connotations of Greek culture, but why not pick a Roman date? The most obvious was the founding of the city of Rome itself. Some have argued that this date could not be used, since historians in Rome could not agree on an exact date for the founding.[8] Such an argument ignores the element of convention involved in the selection of any reference point. Certainly the lack of complete agreement as to the exact date of Christ's birth did not prevent that date from being used. The failure to agree is itself some indication that none of the Roman historians felt strongly the importance of such a date. Moreover, despite the disagreement among scholars, Roman historians dated the founding with confidence in their own works. Both Tacitus and Livy referred to the founding of Rome in precise chronological terms. Livy said that the consular tribunes were established 310 years after Rome was founded,[9] while Tacitus noted the 800th anniversary in his *Annals*.[10]

Since they included a date for the founding of Rome in their narratives, Livy and Tacitus clearly could have used it to count the years, but they preferred a more relative perspective, one which allowed the years to be grouped together in different ways depending on the thematic stress. Livy counted in a number of ways, including the year of the war, especially when covering the war with Carthage. He did not consider the quantitative relationships between the consular years and the war itself to be of great meaning, for on one occasion he referred to two successive consular years as the fourteenth year of the war (28.16,38). Tacitus counted the years of the imperial reign, creating particular sequences of years whose meaning could then be stressed by rhetorical devices. He quantified particular temporal relationships only for dramatic effect and to stress thematic connections between events. At one point, for instance, he observed that Brutus' sister Junia died in the sixty-fourth year after the battle of Philippi (3.76).

It is all too easy to miss the effect of narrating by consular years. Modern editors of these histories have carefully correlated the annual events with our absolute time line. Most editions of classical historians contain these absolute references, so that the reader can easily tell that the events narrated under the consulate of Gnaeus Fulvius Centumalus and Publius Sulpicius Galba, for instance, occurred in 211 B.C., but classical readers did not find such an arithmetical series in their copies, nor could they easily construct one for themselves. Some modern historiographers have suggested that they could look up the date, and in fact lists of officeholders existed in public archives, but the simple truth is that readers could not easily construct a numerical series to locate the events. Except for such particular works as that of Polybius, where the elections for a brief period were correlated with the Olympiad, there was no serial list readily available to readers. The events in Livy and Tacitus had a quantifiable temporal relationship to one another only when the writers themselves ascribed one. No linear number series underlay the events of either story; the chronological order had no intrinsic meaning because it had no existence apart from the events. Since the annalistic form did not aim at a single arithmetic series, it was tied to the events it described in a variety of ways. To understand the time created by this form, the way historians used it must be explored in some detail.

Several factors combined to make the choice of consular year particularly appropriate to Roman historians. First, they had access to lists of officeholders going back to the earliest years of the Roman Republic, lists that included priests and other ceremonial officials, as well as the consuls. These lists, though not strictly reliable before the third century, were accurate from that period to the end of the republic. There was thus a documented, continuous tradition that offered a framework around which to organize events year by year. Second, as Rome assumed dominance in the Mediterranean world, its political lists became more suited to expressing the story of that world than had been the case when Polybius was writing. Third, to the Romans themselves their lists of officeholders seemed a natural way of organizing the events of history. In the Greek world, even late into the Roman period, historians used the Olympiad or the lists of Greek officeholders.

For Livy, who wrote of the Roman Republic when the consuls were functioning heads of state, the consular years were not abstracions; they were years in which two particular consuls dominated the political scene, leading armies, making policy, and settling disputes in Rome. Livy narrated the accession of new consuls in ways that acknowledged the intrinsic reality of that event, even when he had to ignore the beginning of the "consular year." When narrating a series of events at the beginning of 199 B.C., for instance, he continued to refer to Sulpicius and Aurelius, the last year's consuls, as "consul," partly in order to follow each consul's activities without interruption (31.33,37–40,47). Only after completing the account of their activities did he mention the consular elections, which had already occurred when many of the events just narrated had happened (32.1). He acknowledged the passage of astronomical time in this section by noting the occurrence of the autumnal equinox (31.47), but the discontinuity between that time and the political time of the consular elections allowed both sequences to proceed tied to their respective events without confusion or inappropriateness.

Because of the importance of the consuls, considerations of temporal location interact in Livy's work with more concrete issues. For instance, early in the war with Carthage the historian addressed the confusion in his sources over when the siege of Saguntum took place.

> Some have recorded that Saguntum was taken in the eighth month from the beginning of the siege. . . . If this is so, it cannot have been the case that Publius Cornelius and Tiberius Sempronius were the consuls to whom the Saguntine envoys were dispatched in the beginning of the siege, and who in their own year of office fought with Hannibal. . . . Either all these things took up somewhat less time or Saguntum was not first besieged, but finally captured, at the outset of the year which had Cornelius and Sempronius as consuls. The battle off the Trebia cannot be put as late as the consulships of Gnaeus Servilius and Gaius Flaminius, because Flaminius entered office at Ariminum and was elected under the auspices of Sempronius who had gone to Rome to hold the elections after the battle of the Trebia, and had then rejoined his troops in their winter quarters. (21.15)

The concern for dates here is inextricably linked to the question of who was involved in the events being narrated. Livy was not trying to locate these events on an abstract time line; he was seeking to associate them with the correct consuls.

When he treated the early period of Roman history, Livy made the order of years depend even more explicitly on the identification of the consuls. He explained that it was so hard to determine when the battle of Lake Regillus was fought because "One is involved in so many uncertainties regarding dates by the varying order of the magistrates in different lists that it is impossible to make out which consuls followed which, or what was done in each particular year, when not only events but even authorities are so shrouded in antiquity" (2.29).

When mentioning events in which the consuls were not so directly involved, Livy was less concerned with discrepancies in his sources. He placed the first stage performances at Megalesia in two separate years (194 and 191 B.C.) without bothering to tell his reader that there was dispute over the date (34.54 and 36.36). Such an error cannot be ascribed to simple carelessness or to his method of following one source at a time. Recent studies of Livy have shown that he devoted considerable care and thought to the organization and structure of his work and that he read and correlated all the available sources before constructing his narrative.[11] Livy placed the same event in two different years not because he was careless but because he did not conceive of the consular years as an abstract linear series. To him the years were concrete entities and could not be separated from the consuls whose elections signaled their beginning and end. The relation of consular years to one another depended entirely on the thematic unity of the events themselves. If there was no thematic reason to cross annual boundaries, then his consciousness and attention did not do so.

Tacitus wrote of the early empire, when the consuls' role was much diminished, but he tried whenever he could to present the consular elections as real events rather than empty measures of time. He noted especially when the emperor himself was elected consul and, to give further content to the event, compared his consulships one with the other (2.53;3.31). Early in the work, describing the period when consular elections still had some importance, each book usually began with the elections, although they were mentioned only in passing, and Tacitus rapidly moved on to more important matters. On one occasion he concluded the account of a consular year with a thorough description of the emperor's manipulation of the elections to produce consuls who would be favorable to him (1.81).

Livy and Tacitus looked on the consular year as a real group of events and not an arbitrary unit of time. Consequently they did not necessarily organize the events within each year in chronological order. Though Livy tried to indicate the passage of seasons in the interest of clarity, linear order gave place to other considerations in organizing events within the year.[12] The Roman annals which Livy used as a source had a traditional order. They started the new year by telling of the ceremonies surrounding the election of annual offices and then went on to the assignment of provinces, the division of military forces, prodigies, and the reception of foreign embassies by the Senate. This order was only partly chronological.

Because it was thematic, it invited the historian to pursue stories associated with the topics to their ends, even if it meant crossing into the next year or returning to an earlier period in the year in order to begin a new topic.

Tacitus made especially good use of the flexibility inherent in the annalistic form. Though he occasionally began years in the traditional pattern with ceremonial observances (1.54;2.41;4.46;14.20), he preferred to begin the narrative of each year with an event that would stress the overall significance of the year. At the end of the *Annals*, in his account of the rise and fall of the Emperor Nero, Tacitus introduced the years with events that demonstrated the general shape of the emperor's career and made his readers think of the story as a whole. Years began with his mother's marriage to the Emperor Claudius that opened the way to Nero's eventual succession to the imperial title, his adoption as Claudius' son and entry into manhood, and his marriage to the emperor's daughter (12.5,25,41,58). Tacitus introduced the consulship that marked the last year of Claudius' reign with prodigies that foreshadowed a change for the worse (12.64). The consulships of Nero's reign began with his political activity and built through increasingly erratic behavior to the murder of his mother (14.1).

Tacitus seldom specified the temporal relationship between these events that he used to introduce the years and the consular elections that officially began it. Most commonly he used phrases like "sub idem tempus" or "interim," both meaning "at about the same time." In other passages he expressed the relationship among events within the year with the ablative absolute, a grammatical construction that specifies a relationship which is not necessarily one of serial temporal order. Though his phrase is often translated in the sense that "these things happened right after the elections," no such relationship is implied. When Tacitus wished to specify a temporal order, he did so explicitly. For instance, when Tiberius withdrew from Rome at the beginning of his second consulship, Tacitus said that the withdrawal occurred "in the beginning of that year [eius anni principio]," and he clearly marked the end of the year as well (3.31;3.49). His location of Tiberius' departure contrasts strongly with his description of other events in the year, which are located only with phrases like "interim" (3.42).

In the works of Livy and Tacitus thematic and spatial units defined the times, even though the narrative was organized year by year. This fact is clearest in those passages where they tried to synchronize events in different parts of the Mediterranean. The difficulties of synchronization were partly technical, arising from the fact that Roman historians had to work with both Roman and Greek sources, which used different dates to start the year. These difficulties could be surmounted. Livy was aware of the obstacles to translating the dates in his sources into modern ones and made efforts to coordinate the various dates in his Roman sources. He knew, for instance, that during the Punic wars the consular year began on the Ides of March and not on January 1, though he did not seem to have known that, due to the omission of intercalary days in the 190s, the Ides of March during the third century occurred in January (Luce, *Livy*, p. 59).

Even when these technical problems were resolved, events in the East did not fit into the same time as those in the West. They seem to be locked into separate compartments with the different times related to one another in an orderly way. Temporal sequences in both worlds are portrayed as they would look to the Romans; that is, events in the East move more rapidly than they do in the West or than they would appear to move to an Eastern observer. To describe the consulate of Lucius Quinctius and Gnaeus Domitius (192 B.C.) Livy began with their election, outlined the apportionment of provinces, and discussed some preliminary military encounters that occurred as the consuls set off into their allotted territories. Abruptly he interrupted this narrative by noting that it was about this time that the mission sent to the East returned to Rome. This mission was of great interest to the Romans, for "at that time the wars which were going on [in the West] caused less concern to the Fathers than the anticipation of the war with Antiochus [King of Syria] which had not yet begun" (35.23). After a long digression explaining the course of the mission and the antecedent developments that caused it to fail, Livy returned to the narrative of Western affairs and finished out the consulate.

Livy knew that he had not blended the two spheres well, for he apologized to his reader. "I have been driven out of my course, so to say, by blending events in Greece with those in Rome, not because they were worth the effort of recording them but because they were the origins of the war with Antioch. When the consuls were elected—for this was the point at which I turned aside—[they] . . . departed to their provinces" (35.40). From this point Livy took only a few lines to tell us that the year was almost at an end and to prepare us for new elections. Many factors are at work here, including a rich source on Eastern affairs, poor material on the West, and a relatively insignificant year. But underlying all of these is Livy's assumption—certainly shared by his readers—that the year contained two separate time sequences, each deriving its order from the perspective of Roman observers. One, the Western one, contained few events to fill the year; the other, the Eastern, is to modern eyes much longer in absolute time, but in the relative time of Livy it "fit" easily within the year, because it appeared that way to Western eyes.

On occasion we can quantify the difference in temporal sequences. Tacitus, when he turned to Eastern affairs during the consulate of Claudius and Vitellius (47 A.D.) introduced his account with the usual phrase "sub idem tempus" and, after finishing with the east, returned to the West with an unambiguous statement of temporal location, "Under the same consulate, 800 years from the founding of Rome" (11.8–11). Yet the account of affairs in the east, despite this precise location in that consulate, covered a six-year period (from 42 to 48 A.D.) in which Tacitus followed the struggles for the throne of Parthia. He presented the story as a unit whose time was shortened. At one point he even proposed that one of the contenders for the throne traveled 350 miles with an army in two days. From his perspective in Rome, Tacitus looked at the affairs of the East as if they were moving faster than they would had he been observing them in Parthia.

At the end of his account of that year Tacitus mentioned that the Parthians sent a secret mission to Rome to seek help against a cruel tyrant who had seized the throne. As the narrative progressed into other consulates and the mission came closer to Rome, it lost the speed which characterized other events in Parthia. Two years had elapsed in his account when Tacitus again spoke of this mission, as it arrived simultaneously with events occurring during the consulate of Gaius Pompeius and Quintus Veranius (49 A.D.), having thus taken two years to traverse the space between Parthia and Rome. As the embassy entered the perspective of the Romans, it took on the dimensions of Roman time, slowed down, and was synchronized with the events of Rome. The phenomenon is similar to that found in Herodotus' history, where participants moved from one time to another as they traveled. Here the linear quality of Hellenistic time throws the change into relief.

Tacitus would certainly not have described like this an event he actually saw. Were he to see a horseman at a great distance, apparently moving slowly over a short distance, he would instantly realize that the distance was much greater than it appeared, and that the estimate of speed and time would need appropriate adjustment. Such an adjustment would be nearly automatic, as it is with us. But without the preconception of universal and linear time to create a grid against which to measure spatial movement from any distance, his "mind's" eye does not make the automatic adjustment that his physical eye would. Instead he seems to have made precisely the opposite adjustment from that he would have made for actual sight. It is as if the infrequent and slow communications from the East, causing reports of events to be transmitted at larger temporal intervals than for affairs in the West, had imposed on the information the effect of time-lapse photography. Just as this process makes flowers seem to open at great speed, so the interruption and delay of communications made armies seem to move unusually fast. Without the influence of a notion of absolute time, both Tacitus and Livy recorded the events they found in their sources as they actually appeared, without distorting them to fit an abstract and preconceived conceptual structure. The narrator's consciousness was a more direct reflection of the order of events, and the years displayed the same overlapping of temporal realities that characterizes consciousness itself.

Tacitus conveyed the complexity of temporal structures particularly well, using the framework of the consular year to express a variety of times. One of the most dramatic periods in the *Annals* is that surrounding the exposure of Sejanus' conspiracy against the royal house. The story, which Tacitus narrated in fifteen chapters, is largely contained within the single consular year of S. Galba and L. Sulla, though some themes are carried over into the following year, when P. Fabius and L. Vitellius were consuls (33 and 34 A.D.). The period was a fearful one for the Roman upper classes. The last of the supporters of Sejanus were mercilessly hunted down and killed; the emperor's own son was starved to death and his daughter-in-law driven to suicide. To all of these calamities were added economic woes, as the growing burden of private debt led to reform of the debt laws and the near bankruptcy of the economy. Tacitus is vivid in his tale of horror.

> And as the executions had whetted his [Tiberius'] appetite, he gave orders for all persons in custody on the charge of complicity with Sejanus to be killed. On the ground lay the huge hecatomb of victims: either sex, every age; the famous, the obscure; scattered or piled in mounds. Nor was it permitted to relatives or friends to stand near, to weep over them, or even to view them too long; but a cordon of sentries, with eyes for each beholder's sorrow, escorted the rotting carcasses, as they were dragged to the Tiber, there to float with the current or drift to the bank, with none to commit them to the flames or touch them. The ties of our common humanity had been dissolved by the force of terror; and before each advance of cruelty compassion receded. (6.19)

To tell this story Tacitus used, within the framework of single year, four major temporal perspectives. The first organized the actions of the emperor himself. This sequence contains the direct political aspects of the temporal order, since Tiberius fulfilled functions like those of the consuls as he acted in his official capacity. In that institutional role he selected husbands for his granddaughters. Tacitus gave brief biographies of the prospective bridegrooms before turning to a more significant act, the request for an armed bodyguard should he ever attend meetings of the Senate again. By telling of these events Tacitus charted the growing personalization of power and the corruption of republican institutions and safeguards that were among his most important political themes.

In a second temporal sequence of this year, Tacitus looked at the underlying social and economic relations that accompanied these political and institutional changes, especially the problem of debt. Here Tacitus took a broader perspective. Before mentioning the relevant event that occurred in that year—a report to the Senate on debt reform—he gave a brief history of debt laws going back to the earliest period of Roman history. Then he explained the law that was passed during the year, describing the disastrous consequences of a sudden change in regulations concerning debt, and the emperor's attempts to mitigate the effects of the change, efforts that led to emasculation of the reform (6.17). Here Tacitus presented a temporal sequence that moved the reader's attention from a concrete event that could be located in the year to a process as old as Rome itself. The story of property and debt had its own rhythms; changes occurred slowly and within limits that tended to bring bold measures and dramatic innovations to naught.

Leaving the underlying economic sequence, Tacitus returned to the hurried time of political events, as he took up slaughter and recriminations that marked the period. "The old fears now returned with the indictment for treason of Considius Proculus; who while celebrating his birthday without a qualm, was swept off to the senate-house and in the same moment condemned and executed" (6.18). After listing the executions and commenting in the passage quoted above on how they had degraded the popular attitude, Tacitus invoked a third temporal perspective for the events of this terrible year. "I cannot omit the prophecy of Tiberius with regard to Servius Galba, then consul. . . . 'Thou, too, Galba shalt one day have thy taste of empire'" (6.20). With this Tacitus embarked on a digression which told of Tiberius' long-standing interest in prophecy and concluded with Tacitus' own opinion. He pointed out the tricks dishonest prophets used to convince people of their skill but admitted the possibility that there is some overall pattern to human life (6.22). Galba will overthrow Nero to end the Julio-Claudian house and put an end—albeit a brief one—to this tale of woe. The temporal relation of this theme to the year lies in the fact that Galba was then consul—a superficial and coincidental connection—but Tacitus made use of it to introduce a perspective which looked beyond the immediate disasters to a temporal rhythm which restores balance and brings back earlier states.

After the discussion of prophecy, Tacitus introduced the fourth major temporal sequence of the year, one created not by the concrete events but by the consciousness of the participants. "Under the same consulate, the death of Asinius Gallus became common knowledge" (6.23). From the consciousness of this death he passed to more important ones, describing the rumors surrounding the death of Tiberius' son, and finally the news of Agrippina's death. This series of deaths, which occurred at different times but entered the consciousness of Romans in this year, was given precise temporal connection only when Tacitus cited the words of the emperor, who remarked that Agrippina had died two years to the day from Sejanus.

In his account of this year Tiberius introduced four distinct but related themes, each existing in its own temporal field. First, there was the concrete political and institutional sequence tied to the acts of officeholders, the linear and measurable time of the Hellenistic Oecumene. Second, there was the field of underlying social relations which gave meaning to the change in debt laws and which suggested a pattern of debt and inefficacious reform in which little of importance changed beneath the surface pattern of individual greed and bankruptcy. Third, there was the rise and fall in imperial houses, not subject to precise prediction but having an order which those sufficiently skilled can see. Fourth, there was the order of consciousness, existing in the present and thus appropriate to be chronicled in the year, but moving backwards and forwards and, in this year, portraying the rapid demise of the Julio-Claudian house.

Finally the horrible year is over. As if to relieve the reader before plunging again into the tale of massacre at Rome, Tacitus introduced a new topic. After naming the consuls to mark the passage into a new year, he noted that this year the Phoenix appeared in Egypt. Mention of the bird leads him into a digression on the length of its life and the periods that elapse between appearances. Clearly skeptical about much of the legend, he concluded, "The details are uncertain and heightened by fable; but that the bird occasionally appears in Egypt is unquestioned" (6.28). Tacitus introduced this topic out of the normal topical order. As a portent the Phoenix belonged at the end of his account of the year. By telling it here, Tacitus brought these longer temporal sequences more forcefully to his readers' attention and made them part of a perspective that transcends all of the concrete events of the period. After accomplishing this goal, Tacitus returned to the tale of devastation. "But at Rome the carnage proceeded without a break" (6.29).

The annalistic histories presented an extraordinary variety of temporal perspectives. The narratives are strictly organized around a succession of annual units, permitting the simple location of concrete events and the construction of a linear series from these events when the narrator thought it appropriate. But the annalistic form was not limited to events that could be unequivocally put on a time line. These concrete events were artfully blended with less concrete realities, including consciousness, prophecies, and broad historical trends, to convey the complex strata of time that govern the historical substance the historians sought to portray.

CHRONOLOGICAL TABLES

Livy and Tacitus wrote without the aid of a time line because they preferred to, not because they had none available. Eratosthenes' table was the beginning of a tradition of such lists which developed during the Hellenistic period. Among the early successors to Eratosthenes was Apollodorus of Athens, who developed a chronology that implied a relationship between the quantitative measure of time and the events themselves, for he saw recurrent forty-year periods in cultural achievements as pupils succeeded masters. Castor of Rhodes drew up a chronicle from the Assyrians to 60 B.C. Basing his work on that of Apollodorus, Castor assembled lists of the kings of Assyria, Media, Lydia, Persia, Macedonia, the archons of Athens and priests of Sicyon, and he placed them side by side so that readers could use the table to synchronize events in different kingdoms. A century later Claudius Ptolemeus constructed a similar list. There too the various dynasties were set next to one another with no single dating scheme to unite them.

These chronological tables lacked a single, continuous measure of time partly because none of the classical kingdoms had lasted throughout the period covered by all of the others. The unification of the Mediterranean eventually brought Hellenistic culture into contact with a historical tradition of sufficient antiquity to serve as an all-embracing reference: the history of the Jews as recorded in the Bible. The Scriptures offered a unique continuity beginning with Adam and proceeding by generation through the patriarchs to the judges and on through the kings of Israel, whose reigns could be synchronized with the events of classical history. Since the book of Genesis recorded the ages of the patriarchs at the birth of their successors, and the historical books of the Old Testament quantified the reigns of the judges and kings, a single numerical time line could be constructed.

This continuity was among the most distinguishing features of Hebrew history, more important at the outset than the notion of providence and linear progression,[13] but the continuity was more spiritual than political. God's promise to Abraham that He would cause his descendants to multiply in return for obedience to His laws gave continuity to the long history of the Hebrew people. The political accounts of Hebrew rulers went back little further than the earliest dates which classical historians could be sure of. When the Hebrew historian Josephus described the Roman destruction of the Temple at Jerusalem, he dated it with great precision—albeit inaccurately by modern standards—1,130 years, 7 months, 15 days from its construction by Solomon, 639 years, 45 days from its rebuilding by Cyrus, and in the second year of Vespasian.[14] This provided synchronization between Hebrew and Roman history but the period of time it encompassed was no longer than that begun by the date Eratosthenes established for the fall of Troy. Thus to Hellenistic scholars, who wanted to use a dating system to locate and synchronize specific concrete political events and were not so concerned with its mathematical continuity and scope, the time line that could be drawn from the Bible was not so clearly serviceable for a universal one as it might be to us.

The Jewish tradition working among early Christian historians did eventually produce chronological tables that displayed the various chronologies of the classical world both in relation to one another and to a single sequence. The most important of these were the chronological tables drawn up by Eusebius of Caesarea, a fourth-century bishop who also wrote a history of the early church. (Eusebius based his tables partly on the late third-century chronology of Sextus Julius Africanus.) Eusebius' *Chronological Canons*, translated from the original Greek into Latin by St. Jerome,[15] became the basic chronological reference work in the Western world until it was superseded in the seventeenth century. Even then, with improvements by Joseph Scaliger and others, who integrated the various calendar systems into the Julian calendar, Eusebius' work in its broad outlines formed the basis of most historical writing on the ancient world.

Eusebius constructed his tables in vertical columns so arranged that his readers could synchronize the various events of the past with one another. On the far left of the page, in a sequence which continued throughout the work, he placed at ten-year intervals a figure representing the number of years from the birth of Abraham. Alongside this column, a series of columns listed the names and monarchs of the various countries. The number of columns ranged from three to nine depending on how many independent countries there were at the time indicated. The continuity of the major empires was expressed in the left column, where Assyrian, Persian, and Roman empires succeeded one another.[16] Beginning with the first Olympiad and extending to the end of the table, he listed the Olympiad number. In blank spaces between the columns important events were recorded. The chronicler wrote major events like the fall of Troy across the whole page in such a way as to interrupt the columns of numbers.[17]

Though the arrangement of the tables lends itself readily to the creation of a single time line—whether based on Olympiads or on years from Abraham—Eusebius was more interested in a particular synchronization, that between the sacred history of the Hebrews and the profane history of the world's empires. By synchronizing the two sequences Eusebius hoped to show the relative antiquity of Hebrew history.[18] To this end he created certain epochal dates where secular and sacred history intersected, such as the second year of Darius, when the Second Temple was built, or the fifteenth year of the reign of Tiberius, when Jesus started his ministry. The dates he chose for the synchronization were epochal and thematic rather than absolute, and Eusebius willingly fit the chronology to his epochs. For example, he moved Aristogiton's attempt to overthrow the tyranny at Athens forward in time so that it coincided with the building of the Temple, and together the two events introduced a new epoch with the liberation of Israel and Athens. The specific chronologies in the *Chronological Canons* contain so many inconsistencies and chronological errors that one scholar even argued that Eusebius had never written them up in a form to allow synchronization but instead had simply compiled them as separate regnal lists.[19] This argument cannot stand critical scrutiny; Eusebius' inconsistencies can usually be traced to his use of a variety of sources. Despite his interest in synchronizing certain events, however, his very indifference to reconciling the discrepancies in his material serves to underline the problems that confront anyone trying to use Eusebius' tables as the basis for an absolute time line.

Eusebius' own use of the tables gives us some insight into his assumptions about time. He compiled a summary of the events of world history, the *Chronicle*, as a preface to his tables, though he did not correlate carefully the events listed there with the tables. In the *Chronicle* he did calculate the total number of years between Troy and his own day, probably on the strength of Eratosthenes' work,[20] but he did not depend on one dating system to locate the events. Instead he grouped events according to reigning monarchs, and in fact organized around kings so exclusively that he missed the Peloponnesian War.

Nor did he use an absolute dating system in his *History of the Church*, which covers the period from Christ to Constantine. He certainly professed there a deep interest in linear time, citing as one of his main topics "the lines of succession of the holy apostles and the periods that have elapsed from our Savior's time to our own . . . the names and dates of [the heretics and] . . . the martyrdoms of later days down to my own time."[21] To present these four topics, of which only the last is intrinsically episodic, Eusebius used a variety of dates. He measured the lines of succession from the holy apostles partly by telling how long each bishop had reigned when his successor assumed the see and occasionally by noting also the year of the emperor's reign.[22] Nor did he date the martyrdoms and heretics with absolute dates. He located the martyrdom of Polycarp, whom he considered one of the most important bishops and to whom he devoted a lengthy passage, only some time in the reigns of Antoninus Pius and Marcus Antonius (4.14–15).

Eusebius did refer to his chronological tables in the *History of the Church* and on occasion used them to correlate several dates. The most important example of this practice is his attempt to date the birth of Christ. "It was the forty-second year of Augustus' reign, and the twenty-eighth after the subjugation of Egypt and the deaths of Antony and Cleopatra, the last of the Ptolemaic rulers of Egypt . . . while Quirinius was governor of Syria" (1.5). In other sections he used similar temporal correlations to establish the validity of other major events of Christ's life, including the passion (1.9).[23] Though the events of Christ's life are dated in the political sequence of officeholders, his mission is dated in reference to the high priests of Jerusalem, thus making it part of the spiritual sequence of God's promise to Abraham (1.9).

The principal significance of the *Chronological Canons* for later historians lay in Eusebius' attempt to synchronize the spiritual sequence of Hebrew history with the political sequence of Hellenistic history. The theological and philosophical issues implicit in that relationship were profound and required centuries to clarify. In addressing these conceptual problems, thinkers of the Middle Ages found Eusebius' work of much practical use, and his dates were widely used. Like Eusebius himself, however, later scholars tended to see in his tables a collection of separate monarchical lists put beside a list marking the spiritual time from Abraham. These two times remained separate in both a technical chronological sense and in a conceptual sense. In fact the spiritual and political sequences remain separated in the narratives of Western historians through the Renaissance. Not until the the sixteenth and seventeenth centuries, when the foundations of the secular history that abolished the distinction were laid, did scholars develop a single dating system that could be used for both spiritual and secular history.

THE TRANSLATIO IMPERII

Polybius was not alone among Hellenistic historians in his search for a syn-optical view that would present all historical events as part of a single pro-cess, but those who shared his quest did not also share his interest in a single chronology that would express the unity of this process. One of the world historians of that epoch who most deeply influenced later writers was himself almost completely indifferent to dates. Pompeius Trogus lived in first-century Rome and wrote a universal history in Latin called *Histo-riae Philippicae*, conceived in forty-four books. Although the original has been completely lost, about one-tenth of it was epitomized by a fourth-century figure named Justin, of whom we know nothing besides the fact that he condensed Pompeius' work. It might seem dangerous to draw con-clusions from such a condensation, but Justin did not rewrite the work in any major way. Instead he kept closely to Pompeius' form and content, skipping over whole sections of the original in their entirety. Because of the way Justin epitomized Pompeius, he is especially reliable as a conveyer of the earlier historian's chronology. [24]

Justin prefaced his epitome with a description of Pompeius' intentions as well as a brief apology for his own work. There he praised Pompeius for writing a comprehensive history of the world, especially of the Eastern part, so that Romans would know as much about the East as the Greeks knew about the Romans. Justin described the work as a true universal his-tory, containing "the deeds of all ages, kings, nations, and peoples." [25] To write this history, according to Justin, Pompeius put together the works of the Greek historians who wrote their separate subjects, "omitting what was useless, dividing the rest into times [temporibus] and arranging it into a series of events [serie rerum digesta]."

Pompeius linked the various separate histories together with his concept of "translatio imperii," or transfer of empire. He argued that in the begin-ning all nations lived under kings who were primarily concerned with simply defending themselves. Though short wars broke out for specific rea-sons, there was no long-term aggression until Ninus, the king of Assyria, driven by a "new desire for empire," changed irrevocably the course of world history and began a program of conquest which ultimately led to the Roman Empire of Pompeius' own day. History essentially began with this imperium, and Pompeius told its story as it moved from one people or re-gion to another. Each major change appeared as a translatio imperii. When Sardanapolis, the last Assyrian king, was defeated and killed by the Mede Arbactus, Pompeius observed, "This man transferred the empire from the Assyrians to the Medes" (1.3). Near the end of the work, assessing the state of the world in his own times, he said that the Parthians held the empire in the east as if they divided the world with the Romans (41.1).

Pompeius' vision went far beyond Polybius' in both time and space. Polybius integrated affairs in the eastern Mediterranean with those of Rome and Carthage to present world history as a unit beginning with the 140th Olympiad, a period of less than a century. Pompeius saw world history as a unit beginning with Ninus and extending to his own time. Where Polybius had integrated a few decades into a single historical process, Pompeius brought several centuries into one synoptical view. Moreover, Pompeius identified the nature of this unity more broadly than Polybius. It transcended any particular conquest and consisted in conquest itself, a condition that occurred in time and was transmitted from one polity to another.

Since the linear succession of empires dominated the work, one might expect Pompeius to have some interest in a linear, comprehensive dating system, but few major historians have been so indifferent to the chronological location of the events they were describing. The lack of absolute dates is so striking that his twentieth-century editor has protested in frustration that Pompeius' events "hover in a time of fragmented shreds in the unending chaos of eternity."[26] Several features of Pompeius' narrative contribute to this judgment—the complete lack of a fixed point from which to synchronize the various cultures, the vagueness about durations at key stages of the process, the tendency to use geography rather than time as an organizing principle, and the willingness to pass back and forth chronologically without directly informing his reader that he has done so.

Pompeius knew of the existence of absolute dates which his readers would recognize, for he dated the building of Tyre as one year before the destruction of Troy (18.3). But he did not use these dates to describe his major theme. To express and trace the translatio imperii vague references to "meanwhile" or "about the same time" were sufficient. When he came to the 140th Olympiad, which Polybius had long since fixed as a crucial date in the growth of Roman power, Pompeius left the events in that year as unrelated as if they had no particular temporal location. "At about the same time there was a change of succession giving new kings to almost all the empires in the world" (29.1).

The translatio imperii was the basic underlying force and substance of history, not something to be explained by deeper causes. Unlike Polybius, who found the reasons for Rome's rise in its constitution and the character of its citizens, Pompeius saw it as inexplicable. "The Roman fortune conquered the Macedonian" (30.4). The translatio was an event, occurring at a point in time, not simply an example of underlying conceptual principles. Since it constituted a linear process with a definite beginning and course, Pompeius did consider the temporal aspects of the translatio important, and he made sure his readers saw both the duration of the process and the specific points at which major changes of empire occurred. To convey the duration and points of translatio, however, Pompeius did not use abstract dates. The point at which the imperium began was the beginning of its own time and was related to no other event. The work thus opened "in the beginning" with the conquest of Ninus. As imperium moved from one place to another, Pompeius identified the point precisely but always in reference to other events, never by an absolute date. When Rome invaded Greece, he carefully fixed several events in the same year: the Greek embassy to Rome asking help against Philip of Macedon, an earthquake which created a new island in the Aegean Sea, and a simultaneous earthquake in Asia which shattered Rhodes. He stressed that the two earthquakes occurred on the same day, making clear the reason for his precision by saying that the prophets foretold from these earthquakes the rise of the Roman empire to swallow the ancient ones of the Greeks and the Macedonians (30.3–4). He correlated the prodigies with each other and with the Roman invasion of Greece in order to show the significance of the invasion as part of his main theme, the translatio imperii. The correlation was important; the absolute date was not.

His interest in such precise synchronization did not apply to the origins and early conquests of peoples that later held the imperium. He introduced Macedon without referring to any events that the reader would be familiar with. In the midst of his narrative of early Macedonian history, he casually observed that these things were going on while the Scythians were repulsing Darius the King of the Persians (7.1–3). Carthage, too, had no specific temporal relation to other powers until it began to expand (19.1).

Duration too was of thematic interest, but Pompeius felt no need to quantify it in absolute terms. The first indication of time in the work occurs after the death of Ninus, when we learn that he reigned forty-two years. At the end of his dynasty, we are told, the Assyrians held their empire for 1,300 years (1.2). This interest in duration was his fundamental chronological concern. He cited the length of Median rule, of Cyrus' reign, of Pisistratus' rule in Athens, and of Philip of Macedon's reign (1.6,8;2.8;9.8). At major junctures he summarized these particular figures in order to stress the passage of time and the duration of important empires. After the fall of Macedon he recorded the number of kings, the length of their cumulative reigns, and the total time that Macedon had held the imperium (33.2). But he was not at all interested in giving these durations systematically enough for the reader to be able to count the years from the beginning of imperium through to its end.

He used geography more consistently than time to organize the narrative, as imperium moved from one part of the Mediterranean to another. Justin even referred to Pompeius as a traveler at one point, saying "having finished the affairs of the Parthians and other Eastern countries and of almost the whole world, Trogus returns home to relate the rise of the city of Rome" (43.1). Geography in a sense was more adapted to his theme than time, since the antiquity of a particular power had less to do with when it acquired the imperium than did where the power was located, as the empire was gradually transferred from east to west.

The linear sequence, with its own duration and fixed points of major change, that carried the translatio imperii existed side by side in the narrative with other major events that existed in other times and were often narrated episodically. His treatment of both the Scythians and the Carthaginians reveals this overlapping sense of times. The Scythians presented a particular challenge to Pompeius' interpretation, since they antedated the conquests of Ninus and forced him to begin the story of the translatio after the earliest events of history. In Book Two, after the full account of the Assyrian imperium that had begun with Ninus, he turned to discuss the Scythians. At first the reader has no indication that the narrative has in fact gone back in time. We are told of the Scythians' antiquity and of their ability to repel the attacks of such later conquerors as Darius, Cyrus, Alexander the Great, and the Romans. Then we learn of an Egyptian invasion which induced the Scythians themselves to attack Asia, where they stayed for fifteen years and from which they exacted a tribute for fifteen centuries until Ninus ended it (2.3). With the reference to Ninus, Pompeius finally informed the reader that he had in this book moved back to a period of time before the putative beginning of history. The thematic unity of the narrative, the translatio imperii, existed in a time separate from that of any individual polities. It had its own beginning and progress as it moved from Mesopotamia to Rome. Other events which Pompeius thought worth mentioning existed in their own times without immediate connection to the time of the translatio.

Carthage, which never received the imperium, did not enter fully into the linear time by which Pompeius narrated the translatio. Important though its struggle with Rome was as a factor in permitting Rome to assume the imperium from Macedon, Carthage itself and Rome's battles with it appear episodically in the narrative. Pompeius did not attempt to form a linear series out of them. He told of the period of Carthage's ascendancy during the Second Punic War as part of the growth of Macedon. When Pompeius discussed the decision by Philip of Macedon to enter Italy and attack the Romans, he explained that the Macedonian king was convinced by Rome's defeats at Lake Trasimene and Cannae that she was weak enough to be conquered (29.4). In the following book Pompeius told the story of Rome's ascendancy. That book opened with an Egyptian embassy to Rome seeking help against Philip. Only somewhat later in the book, when Pompeius mentioned in passing that Rome after defeating Hannibal was now ready to move against Philip by helping his enemies, does the reader become aware of the end of Rome's war with Carthage. The Second Punic War itself had no status in linear time; it existed only as episodes in the translation of power from Rome to Macedon.

It is possible that Pompeius included more dates than Justin recorded in his epitome, though the absence of absolute dates is too nearly complete to be without significance. In any case Justin was not interested in the dates and felt the story stood without them. Not only that, but the historians who came after him were equally unperturbed by the absence of dates. For a millennium Justin's epitome constituted the basic source for classical history of the east and provided the narrative framework for most of the historians discussed in the next chapter. The translatio imperii served as a potential framework for all secular history and provided the basis from which medieval historians saw history as parallel linear series of spiritual and secular events.

Pompeius was not the only historian to fit events into a larger unity. P. Annius Florus, a contemporary of Hadrian, wrote an epitome of Roman history in which he likened the Roman people to a single individual whose whole life could be reviewed from birth to old age. According to Florus, the first 400 years—the rule of kings—was the infancy of Rome; the next 150—from Brutus to Appius Claudius—the youth; the 150 from Appius Claudius to Augustus, the manhood; and the 200 from Augustus to his own time, the old age, when Rome had lost its potency and vigor.[27] (This division is not new with Florus. It goes back to the elder Seneca, who used it in his histories.)

Like Pompeius, Florus did not rely on precise dates to present Roman history in this schema. He used Livy as a source for most material in the *Epitome*, and he adopted Livy's practice of dating by consulship. Usually the reader does not find even the consulship as an indication of time, and many passages suggest that the author measured his times in different ways according to different purposes. At the mid-point of the work, Florus paused to assess the period of manhood. There he divided it into two 100-year periods, the first period humane, the second given over to cruel internal strife (1.47). Thus manhood from the perspective of the whole is a 150-year period, while within its own time frame it lasted 200 years. Clearly Florus, like Pompeius two centuries before him, looked at large unities as temporally significant due to thematic and internal relations, not as connected with any external quantifiable dating scheme.

The Scope of Linear Time in the Oecumene

The Hellenistic Oecumene changed the face of time. To describe and explain the new integration of world affairs, historians created a time of vastly greater chronological scope than that of their Hellenic predecessors. New dating systems appeared to measure this larger block of time, and new conceptions of time organized and interpreted the events within it. But the innovations did not replace older dating systems and modes of narrative. They were used to understand the unity of the Oecumene, but for other sorts of events older chronologies sufficed. Except for Polybius, historians did not connect the new chronologies of this period with the conceptual unity. Hellenistic historiography thus continued to utilize a variety of time frames, both linear and episodic, to express different historical insights.

The Oecumene did add an important element to those processes that classical historians had included in linear time. Alongside the knowledge of religious matters which Herodotus had put into linear time and the acquisition of technical skills which Thucydides had seen as linear, the growth of empire became a linear process that could be measured and expressed on a numerical scale, even when it passed from one culture to another one that used an entirely different dating scheme. Because political change was now linear, some of the stories handed down from the Greeks began to seem unreal to historians of the Hellenistic period. Pompeius Trogus, in telling of the fall of the false Smerdis, eliminated the part Prexaspes played in Herodotus' story (Pompeius, 1.9). According to Pompeius the conspiracy among the nobles was solely responsible for the downfall. Pompeius essentially saw the story as a political one and converted it into the linear time of the Oecumene.

Political legitimacy had become a linear process, but other historical realities, whose role in the overall story made them indispensable parts of the narrative, remained outside the linear time frame. In particular, the individuals whose decisions and character accounted for so much of the moral value of history were not part of the time in which the translatio imperii occurred, just as they had not formed part of the more limited linear time of Herodotus and Thucydides. Among the Greeks the exclusion of individuals was less noticeable, but because of the Hellenistic accent on individualism historians paid more attention to the role of individuals in the historical process. The resulting narratives displayed eccentricities similar to those noted in Herodotus' narrative of the fall of Smerdis.

The emperors of the Julio-Claudian house, with their sexual exuberance, reckless extravagance, and wanton cruelty, were especially memorable individuals, and some of the most influential narratives of the Hellenistic period depicted their actions and character. Suetonius' *Lives of the Caesars*, as well as the works of Tacitus, both presented impressive accounts of the reigns of these emperors in which historians for centuries have found rich material for moral lessons and compelling anecdotes.

Among the most fascinating and problematic of the Julio-Claudian emperors was Augustus' successor, Tiberius, whose reign covered those decades when imperial government became firmly rooted as the basic political structure of Rome. Modern historians have studied Tiberius' role in making these institutions permanent, but the immense fascination Tiberius has exercised over the centuries is due to his personality and character. The emperor's retirement to Capri gave rise to rumors of sexual escapades and drunken orgies that served as the envy and scandal of future libertines and moralists. The emperor's attacks on the members of his own family and his bloody suppression of revolts gave him an unsurpassed reputation for cruelty and rapaciousness.

But Tiberius was more than this. The young Tiberius was completely different from the cruel and lecherous old emperor. In early life he had a splendid military and administrative career, acquiring a reputation for austerity, restraint, and conscientious performance of his duty. Many explanations of the change in Tiberius' character have been adduced over the years. The most common and convincing to modern ears argues that the influence of imperial power corrupted a figure whose baser tendencies were kept in check during his youth by the restraints of tradition.[28]

Such a picture of Tiberius' life is convincing to us because it places his known behavior in a linear time frame, reconciling contradictions in his character by incorporating the events of his personal life on the same time line that expresses the political changes of empire. But the historians who recorded his life did not integrate the personal and political realities into a single time frame. They understood the development of imperial institutions as a linear process which permanently displaced republican Rome, however much they might disagree on the significance or value of the change. But the personal life of Tiberius existed in another time, one which could encompass contradictions and paradoxes that linear time could not accommodate.

Two separate pictures of Tiberius' character, overlapping in time, emerge from Suetonius' description. In the first half of the biography, discussing his civil and military career down to the middle of his reign as emperor, Suetonius emphasized his sense of responsibility, modesty, and austerity. Tiberius slept in the open and ate simply while on campaign, went to great lengths to live a simple life and avoid the honors due a member of the ruling house during his residency on the island of Rhodes, and treated the senators with elaborate courtesy after succeeding Augustus as emperor.[29] His sexual life was extraordinarily modest. He was so in love with his first wife, whom Augustus forced him to divorce, that when he saw her after the separation he followed her litter in tears (3.7). He disapproved of his second wife for her attempts to seduce him before the marriage and was driven to leave Rome partly by her immorality (3.10). Even so, his sense of familial loyalty and responsibility was so high that he tried to reconcile her with her father after Augustus had finally discovered her conduct and banished her (3.11).

In the middle of his reign Tiberius left Rome for Capri. In telling of this part of his reign, Suetonius introduced the other side of Tiberius' character.

> Then returning to the island he utterly neglected the conduct of state affairs, from that time on never filling the vacancies in the decuries of the knights nor changing the tribunes of the soldiers and prefects or the governors of any of his provinces. . . . Moreover having gained the license of privacy, and being as it were out of the sight of the citizens, he at last gave free reign to all the vices which he had for a long time ill concealed, and of these I shall give a detailed account from the beginning. (3.41–42)

This introduction leads into a remarkable list of weaknesses which Tiberius had shown in his early career, including drunkenness, gluttony, sexual immodesty, extravagance, and irresponsibility in office. Only after listing these early vices, did Suetonius venture to describe the sexual excesses and cruelty that characterized the emperor's behavior at Capri.

Suetonius presented two simultaneous pictures of Tiberius' character. He could easily have incorporated them into a linear story by telling us that the early responsible behavior of the emperor was a sham, a part of his public persona, and a mask assumed purely to hide his true self from the public. But Suetonius gave no hint of Tiberius' duplicity in that early section. Nor can some of his behavior there, such as his grief over the separation from his first wife, easily be explained in those terms. Suetonius and his readers accepted the notion that contradictory characteristics could exist simultaneously in the same person's life, existing in different times according to the use the historian had for events. In this case Suetonius devoted the first part of his biography to showing the virtues of character in the second emperor of Rome; the second part, to showing his private vices. (In treating other emperors, including Augustus, Suetonius adopted the same approach of separate treatment of virtues and vices that existed simultaneously.)

Tacitus' annalistic organization compelled him to place the events of Tiberius' life more clearly on a single time line, but that order did not lead him to a linear picture of Tiberius' character. In his concluding passage on Tiberius, Tacitus placed the emperor's character in several times.

> His character again has its separate epochs [tempora . . . diversa]. There was a noble season in his life and fame while he lived a private citizen or a great official under Augustus; an inscrutable and disingenuous period of hypocritical virtues while Germanicus and Drusus remained; with his mother alive, he was still an amalgam of good and evil; so long as he loved or feared Sejanus, he was loathed for his cruelty but his lust was veiled. Finally when the restraints of shame and fear were gone, and nothing remained but to follow his own bent [ingenio], he plunged impartially into crime and into ignominy. (6.51)

Tacitus here applied the same division between private and public character that Suetonius had used, but Tacitus brought the two characters into a single temporal context. Yet Tacitus' observations are overlapping; they lead the reader to contemplate the emperor's relationship with specific people and influences in his life, and they militate against a single continuous picture of his development. The narrative itself reveals a similar complexity. When Tacitus described each of the turning points mentioned in that passage he saw it as a self-contained event, to be explained in terms that often conflicted with the influences he ascribed to other individuals at various points in the narrative.

The great changes in Tiberius' character—and in the fate of Rome itself that had become so entwined with the emperor's—retain an air of mystery and ambiguity.

> The consulate of Gaius Asinio and Gaius Antistius was to Tiberius the ninth year of public order and domestic felicity (for he

counted the death of Germanicus among his blessings) when sud-
denly fortune disturbed the peace, and he became either a tyrant
himself or the source of power to the tyrannous. The starting
point and the cause were to be found in Aelius Sejanus. . . . Before long by his multifarious arts, he bound Tiberius fast so much
so that a man inscrutable to others became to Sejanus alone un-
guarded and unreserved and this less by subtlety . . . than by the
anger of heaven against that Roman realm. (4.1)

Tacitus located the change clearly in the proper consulates and in the
ninth year of Tiberius' reign, but he traced the cause both to a specific
influence in that year and to a general turn of inexplicable fate. Tiberius'
character existed both in the single time frame where Tacitus saw the workings of imperial rule and in the multifaceted, overlapping times where different events and circumstances brought out different virtues and vices.

An absolute, linear chronology, though developed to new levels of complexity and comprehensiveness by mathematicians and chroniclers from
Eratosthenes to Eusebius, remained in the eyes of most practicing historians an artificial device, not entirely suitable for telling a story in its totality. In Augustan Rome a literary critic and historian named Dionysius of
Halicarnassus expressed this feeling in a comparison of Herodotus and
Thucydides.[30]

> Thucydides keeps close to the chronological order, Herodotus to
> the natural grouping of events. Thucydides is found to be obscure
> and hard to follow. As naturally many events occur in different
> places in the course of the summer and winter, he leaves half-
> finished his account of one set of affairs and takes other events in
> hand. Naturally we are puzzled, . . . our attention is distracted.
> Herodotus, on the other hand, begins with the dominion of the
> Lydians and comes down to that of Croesus. . . . He relates some
> of the events as a sequel, takes up others as a missing link. . . .
> Although he recounts affairs of Greeks and barbarians which oc-
> curred in the course of some two hundred and twenty years on the
> three continents and finally reaches the story of the flight of
> Xerxes, he does not break the continuity of the narrative. The
> general result is that whereas Thucydides takes a single subject
> and divides one whole into many members, Herodotus has chosen
> a number of subjects which are in no way alike, and has produced
> one harmonious whole.

Dionysius' critique, written at the height of the Hellenistic period, sheds
light on the continuing concern among historians for a natural narrative,

one not bent into a single time frame and forced to follow an arbitrary order. His praise of Herodotus for abandoning linear sequence in order to maintain the continuity of the story could be extended to Tacitus, Livy, or Pompeius Trogus. And it echoes Polybius' own doubts about how readers would receive those sections of his own work where stories are distorted by his Olympiad year. Clearly temporal order for this period depended on the exigencies of particular stories. Thematic unity did not demand that events be represented as part of a strict chronological series.

5

The Time of the Incarnation

What then is time? Who can find a quick and easy answer to that question? Whoever in his mind can grasp the subject well enough to be able to make a statement on it? Yet in our ordinary conversation we use the word "time" more often and familiarly than any other. And certainly we understand what we mean by it, just as we understand what others mean by it when we hear the word from them. What then is time? I know what it is if no one asks me what it is; but if I want to explain it to someone who has asked me, I find that I do not know.

Augustine, Confessions 11.14

THE BIRTH OF PERSONAL TIME

Augustine was forty-three years old when he wrote these words, and he knew well the force and mystery of time, for it had brought many changes to his own life. He had risen in his chosen profession as a teacher of rhetoric to the highest and most prestigious ranks, holding posts in Rome and Milan; he had in his adolescence embarked on a sustained study of classical philosophy, beginning with Cicero's summaries of major Hellenistic schools of thought, moving through the rigid and superstitious dualism of the Manichees, and passing to the books of the Neoplatonists. Less than a decade before the *Confessions* this odyssey had culminated in his conversion to Christianity, his rapid rise to prominence among Christians, and his appointment as a bishop in Africa. In that decade he began the series of major writings that would make him a founder of Christian orthodoxy.[1] Augustine's curiosity about time was not a speculative whim but a deep personal need to understand the changes that had made him who he was.

Because his question arose from a personal need, he found unacceptable those definitions of time that bound it to the motions of objects in the natural world, however exalted those objects might be.[2]

> I once heard a learned man say that what constitutes time is the motions of the sun and moon and stars. I did not agree. For one might equally well say that the motions of all bodies constitute time. Suppose that the lights of the heavenly bodies were to cease and a potter's wheel were to be turning around: would there be no time by which we could measure its rotations and say that these rotations were of equal duration, or, if it turned sometimes faster and sometimes slower, that in the one case the period taken by the turn was shorter and in the other case longer? . . . There are stars and lights in the heavens to be for signs, and for seasons, and for years and for days; there certainly are; yet, just as I should not say that one turn of that little wooden wheel constituted a day, so that learned man should not say that it does not constitute any time at all.

Not only is time different from any particular natural movement, but even motion itself does not constitute the source of time.

> For when a body is in motion, I measure the length of the motion in time from the moment when it begins to move until the moment when it ceases to move. And if I did not observe the moment when the movement began and if the movement continues to go on so that I cannot observe the moment when it ends, then I am incapable of measuring it. . . . It is clear, then, that the motion of a body is one thing and the means by which we measure the duration of that motion is another thing. Is it not obvious which of the two deserves the name of "time"? A body may sometimes be in motion, at varying speeds, and may sometimes be standing still; but by means of time we measure not only its motion but its rest. . . . Time, therefore, is not the motion of a body. (11.24)

Up to this point in his argument, Augustine was following accepted scientific definitions of time. Aristotle, whose logic Augustine admired and felt so proud to master (4.16), had maintained that time was the quantity of motion (*Physics*, 4.14.223a), and as his Hellenistic commentators worked out the implications of this definition, they made clear that it was meant to separate out the pure measurement of time from the natural motions that were commonly used to measure it. Strato of Lampascus, one of his successors, maintained that days and years were not time nor part of time, but purely the revolutions of the sun and the moon. He felt time was the quantity in which these perceptible revolutions existed.[3] Aristotle even understood that his definition raised the question whether time was independent from the mind. "Whether if the soul did not exist time would exist or not is a question that may fairly be asked; for if there cannot be someone to count there cannot be anything that can be counted" (*Physics*, 4.14.223a). But Aristotle was only marginally interested in the problem. He did not acknowledge so fully as Augustine the reality of the personal soul, and the Greek philosopher willingly accepted the conceptual possibility of unobserved motion, arguing that time was a quantifiable aspect of such motion.

For Augustine such an abstract solution did not suffice. He was sure that time must exist in the soul, for how else could we speak of a past time, since the events that comprised it no longer exist, or of a future time, since its events do not yet exist?

> It is now, however, perfectly clear that neither the future nor the past are in existence, and that it is incorrect to say that there are three times—past, present, and future. Though one might perhaps say: 'There are three times—a present of things past, a present of things present, and a present of things future.' For these three do exist in the mind, and I do not see them anywhere else: the present time of things past is memory; the present time of things present is sight; the present time of things future is expectation. (11.20)

The past, present, and future correspond to three parts of our soul, and they have a unity in that soul which transcends the physical motions of nature.

The word Augustine used to express this unity of the soul in which we measure time was "attentio," attention, or, to use a more modern term, consciousness. It is attention that allows us to make judgments about whether a time is long or short, since the past—or the future—is neither long nor short; only our expectation and memory have length. To explain how the attention works to measure the past he analyzed the process of reciting a psalm he knew.

Before I begin, my expectation (or "looking forward") is extended [tenditur] over the whole psalm. But once I have begun, whatever I pluck off from it and let fall into the past enters the province of my memory (or "looking back at"). So the life of this action of mine is extended [distenditur] in two directions—toward my memory, as regards what I have recited, and toward my expectation, as regards what I am about to recite. But all the time my attention (my "looking at") is present and through it what was future passes [traicitur] on its way to become past. (11.28)

The past is thus a creation of the soul's attention as it converts the contents of expectation into memory, endowing those contents with length out of its own knowledge and experience. Furthermore, it remains true that our conscious attention creates the past no matter how small or how vast the interval. "And what is true of the whole psalm is also true of every part of the psalm and of every syllable in it. The same holds good for any longer action, of which the psalm may be only a part. It is true also of the whole of a man's life, of which all of his actions are parts. And it is true of the whole history of humanity, of which the lives of all men are parts" (11.28).

There are in fact two concepts of time here.[4] From one perspective Augustine made a single temporal series of all human history, a series created by God and transcending any particular order of events. This series is indeed absolute and can be used as the basis of an absolute time line. This aspect gives Augustine's concept of time a quality that seems to look forward to Descartes and Newton and to provide a conceptual framework from which the absolute time of the seventeenth century could emerge.

There is a second perspective here, however, one different from classical models but not consistent with the demands of absolute time. Augustine saw the intense particularity of past events, portraying each in its own temporal framework with a beginning, an end, and a length appropriate to it. From this perspective, duration is not an abstract number but an immediate personal judgment; it depends on the observer. Since expectation and memory determine measurement, the length of each event arises from our attention as we discern from our own expectations and memories the outlines and singularity of a particular event. Thus from this second perspective each event or series of events requires a measure appropriate to it and determines the shape of its own time. Though the first perspective seems more "true" in a Newtonian sense, the second is the one which dominated Augustine's narrative of his own life.

The Confessions, where Augustine laid out this theory of time, demonstrates his use of a dating system arising out of the subject matter. The system he used there depends on the personal soul creating time in the process of measuring it. The work is devoid of absolute dates, even though such important political figures as the prefect of Rome and the bishop of Milan crossed Augustine's path. He could easily have oriented his reader to a broad and impersonal political chronology by connecting the events of his life with the political lists. Certainly Hellenistic and Greek historians, such as Arrian or Plutarch, used dates in this way, even when writing the biographies of individuals.

Writing of his private life, Augustine chose the dating system most appropriate to that subject: his own age. Furthermore he used his age to direct the reader away from a simple numerical chronology and towards the deeper realities Augustine saw in time. He located certain important events in his life with precision—the crisis of his sixteenth year, when he discovered the evil within himself, the revelation of his nineteenth year, when he read Cicero and embarked on a career in pursuit of wisdom, and the disillusionment of his twenty-ninth year, when he met Faustus the Manichaean and realized that the Manichees could never answer his questions. These crises all exist as points on a continuous time line, but in other sections of the work this precise series of points is absent. Whole periods pass without chronological indication, portrayed as a single epoch whose events are not necessarily related in a linear chronology, just as Herodotus had portrayed the periods of Croesus' life. "So for the space of nine years (from my nineteenth to my twenty-eighth year) I lived a life in which I was seduced and seducing, deceived and deceiving, the prey of various desires. My public life was that of a teacher of what are called 'the liberal arts'. In private I went under cover of a false kind of religion. I was arrogant in the one sphere, superstitious in the other and vain and empty from all points of view" (4.1). In such terms did he describe his early youth, expressing the simultaneous quests of wisdom and ambition, the one down the false paths of the Manichees and the other in vain pursuit of his career in rhetoric.

Some of the most important events of his life lack precise quantitative chronological connection with other events; they do not exist as points on a linear time line even though they are concrete events. The very conversion which climaxes the work is not dated in the same manner as some of his earlier crises. Although there is some indication of his age, Augustine made clear that he intended it to be only an approximate one. When discussing his depression just before his conversion, he observed, "Many years (at least [forte] twelve) of my own life had gone by since the time when I was nineteen and was reading Cicero's *Hortensius* and had been fired with an enthusiasm for wisdom" (8.7). The qualifying "forte" seems strange, since he knew and recorded his exact age when he encountered Faustus only four years before. The disclaimer serves to separate his conversion from the temporal sequence that led up to it. The scene in the garden where he heard the voice of a child, read the lines of St. Paul, and found the peace he had longed for, is the climactic scene of the *Confessions*. It certainly depended on the events of Augustine's earlier life, but not because of the number or linear order of the years that had passed. In fact, in that very passage where he told the reader he did not know his exact age, he complained that others had followed shorter paths to conversion than he had. The length of time is not absolutely meaningful; his story consists of specific events whose relation to each other was thematic rather than purely temporal. The conversion was an event that brought together the episodic and linear qualities of time that created his personality.

In the *Confessions* the astronomical units are conveniences for marking the passage of time. Time itself is personal, created by the attention of the soul as it moves events from expectation to memory. This personal time created its own chronology, just as did the time of the Oecumene, a chronology in each case that expressed the unity which underlay it. In dating the events of his life with reference to his own age, Augustine blended two senses of time, one where events appeared as distinct points in a linear series, and one where events had no clear linear relationship to one another and where only the episode as a whole had a place in the linear time of Augustine's personal development. The distinction arose from his conscious perception of the meaning of his life as a whole and was a crucial tool in communicating that perception. Augustine's chronology, like Polybius', related events to one another only when the author perceived the relationship among events as a temporal one.

As long as the subject of Augustine's inquiry into the past was a person, the overall connection with time was clear. Birth and death give personality an unmistakable connection with temporal process. The temporal relations among the events of a person's life have an intrinsic meaning. But Augustine also felt that human history in general had the same relation to time that any particular person's life did; that one could understand the course of events only by comprehending their general progress from expectation to memory. Implementing such a theory on a general scale presented formidable problems. Most important, human history had no physical body; the basic empirical phenomenon which gave assurance of the importance of time for a person's life was absent from the general story of mankind. Some other reality must take its place, less visible than the body but still clear enough to give intrinsic temporal structure to events.

More than a decade after completing his *Confessions*, Augustine addressed himself to the problem in the *City of God*. He began this work in response to a real historical crisis—the sack of Rome by the Visigoths. He wrote the work to defend Christianity from the charge of having caused the calamity by deserting the pagan gods, but his answer to that charge led him into a profoundly influential theory of history. After an initial section in which he showed that such disasters were hardly uncommon in the history of the world, Augustine went on to argue that the sack itself could not be considered an unmitigated evil. In fact the meaning and value of any historical event could be accurately judged only after its place in the whole course of human history from beginning to end had been understood. This argument in turn led Augustine directly into problems of chronology, for in order to understand the course of human history he needed to measure the temporal intervals among significant events and create a chronological framework for assessing the overall pattern.

To describe the pattern of history Augustine took from both biblical and classical sources a schema known as Age theory. He divided the course of world history into a succession of ages counted by the days of creation, which he called the Great Week.[5]

> This sabbath [the eternal rest of the saved at the end of time] shall appear still more clearly if we count the ages as days, in accordance with the periods of time defined in Scripture, for that period will be found to be the seventh. The first age, as the first day, extends from Adam to the deluge; the second from the deluge to Abraham, equaling the first, not in length of time, but in the number of generations, there being ten in each. From Abraham to the advent of Christ there are, as the evangelist Matthew calculates, three periods, in each of which are fourteen generations—one period from Abraham to David, a second from David to the captivity, a third from the captivity to the birth of Christ in the flesh. There are thus five ages in all. The sixth is now passing, and cannot be measured by any number of generations. . . . After this period God shall rest as on the seventh day, when he shall give us (who shall be the seventh day) rest in Himself.

Human history thus displayed the same features of personal growth and development that Augustine himself had manifested in the *Confessions*. Augustine reinforced the parallels by referring to the different ages as stages of an individual life. He called the reign of David a starting point for the youth of the City of God. There were fourteen generations between Abraham and David, he explained, because at fourteen men enter puberty. The other ages of world history also express particular stages of individual life.[6]

> Previously this family of God's people was in its childhood, from Noah to Abraham; and for that reason the first language was then learned, that is, the Hebrew. For man begins to speak in childhood, the age succeeding infancy, which is so termed because then he cannot speak. And that first age is quite drowned in oblivion, just as the first age of the human race was blotted out by the flood; for who is there that can remember his infancy?

The personal metaphor is not incidental to Augustine's treatment of human history. It reflects the deepest levels of his theology, for it comes out of the very doctrine of the incarnation which proved such a stumbling block to his own conversion and which constituted for him the essence of Christianity. If Jesus was God incarnate, then God could be a person, experiencing all the dimensions of life, including birth and death. If God was a person, then time took on the deepest reality. By locating time in his own soul Augustine was not diminishing its significance but enlarging it, for the same memory and expectation which gave him his sense of time was also the repository of faith and hope and of his experience of God. "Since the time I learned you, you stay in my memory and there I find you whenever I call you to mind and delight in you" (*Confessions*, 10.25).

The implications of the incarnation for history and time are vast. Christian thinkers required centuries to apply them to all dimensions of experience and create a historical narrative in which personality plays the full role implied in this doctrine. The incarnation introduced a new reality into the historical process that proved especially hard to accommodate. Will and intentionality became central features of human nature. No longer subordinated to the mind and intellect, desire shaped and molded human destiny with its own dynamic. It constituted for Augustine the distinguishing feature of human nature from birth to death and lay behind such distinctive Augustinian doctrines as original sin. His personal life derived its unity from the will whose affections directed it and were themselves subject to change in the process of growth. History as a whole derived its unity from the Divine will that directed it towards an end which God knew and had partially revealed.

Age theory was quite effective in giving an overall structure to the past and providing a conceptual basis for the Augustinian picture of history as an emanation of God's will. It loomed large in theoretical discussions of history throughout the Middle Ages. It was less useful as a practical tool for constructing a narrative.[7] According to the schema all history since the incarnation was placed in the sixth age, and as that period lengthened, the inadequacy of the division became more apparent. More important from a chronological standpoint, the ages corresponded to no fixed dates. As Augustine himself remarked in the preceding passage, they represented generations rather than years. To make precise temporal correlations among events of the sixth age, medieval historians would need another system.

The *City of God* is not a historical narrative but an essay on the meaning of history, and in it Augustine used historical events only as examples of his theory. Even so he had need of some chronology for his most important arguments. In particular he needed to synthesize the events of secular history into a single process, ruled by God but directed by selfish human desires, and to correlate this process with the religious history whose growth he plotted with Age theory. To accomplish this synthesis Augustine modified the translatio imperii of Pompeius Trogus. He looked at secular history as the story of two major empires, the Babylonian and Roman, which were in reality the incarnations of a single human empire.

> Among the very many kingdoms of the earth into which . . . society is divided (which we call by the general name of the city of this world), we see that two, settled and kept distinct from each other both in time and place, have grown far more famous than the rest, first that of the Assyrians, then that of the Romans. First came one, then the other. The former arose in the east, and, immediately on its close, the latter in the west. I may speak of other kingdoms and other kings as appendages of these. (*City of God*, 18.2)

To establish these temporal relations between Rome and Babylon, Augustine used the *Chronological Canons* of Eusebius but concentrated on the generations rather than the years listed there. Even when he synchronized the religious sequence of Hebrew history with secular history, he still considered only generations. He calculated that Abraham was born when the second kings ruled Assyria and Sicyon and that he departed from Babylon when the seventh king ruled Assyria and the fifth reigned in Sicyon. He was definitely more interested in temporal relations than Eusebius, whose own account of the place of Roman history in God's plan was vague and without chronological references from his own tables (*History of the*

Church, 1.2). But Augustine's interest in the temporal order of events required no absolute dating system; the generational one sufficed for his purposes. In a sense the generational order was more appropriate than the solar year, since it was closer to the personal reality that gave time its meaning. Augustine made abstract temporal judgments only to calculate such things as the relative durations of wars or to determine the possible occurrences of solar eclipses (*City of God*, 5.22–23;3.15).

So unconcerned was Augustine with the temporal progression of secular events from Babylon to Rome that he felt free to ignore events in the East and concentrate on Greek and Roman affairs, which were better known to him than the earlier and more distant ones. He certainly could document western affairs more easily, but there were conceptual issues behind his choice. He was convinced that the religious and secular order could be compared by looking at Rome alone, since Rome was essentially a second Babylon, having taken up rule from the point where Babylon ceased to be (*City of God*, 18.2).

The concept of a single time embracing all human history did not, then, lead Augustine to use a single dating system. In both the *Confessions* and the *City of God* he used relative chronologies that placed a particular event in the context he chose for it. In the *Confessions* he dated some events from his birth but also described large periods of his life when events had no particular linear relationship. He looked at his conversion as belonging to another sequence that did not depend on counting the years from his birth. In the *City of God* he synchronized his two themes by counting generations, even though the chronological tables he was working with would have allowed him to date by solar year with some precision.

But the new time he created did influence Augustine's chronology. It had a personal cast to it that differed from the political dating systems of the Hellenistic world. Those historians used generations because they reflected the importance of the ruler; they changed to the solar year to express political relations that transcended a particular country. Augustine used generations rather than the solar year because the generations expressed more directly the element of personal growth in time itself. Furthermore his connection of Babylonian and Roman history indicated his commitment to a unified time whose significance could be seen by the human mind. He considered the two empires not as the greatest in fact but as the most important in fame. It was in the consciousness of those looking at the past that the empires of the world took on a unity, just as in the *Confessions* he showed that the unity of time itself depended on the creative activity of the human soul. Consciousness was the determining factor in his concept of time, for it connected the external events of the political world with the internal desires and feelings which gave reality to the passage of time. This crucial difference between Augustinian time and the time of the Oecumene was of great potential significance, but it did not immediately affect the development of medieval chronology. The actual dating system which came to dominate medieval historiography came from other roots and bore the imprint of other considerations.

ANNO DOMINI

Gregory of Tours began his *History of the Franks* with a confession of perplexity that seems remote from Augustine's confidence in the meaningful pattern of history. "A great many things keep happening, some of them good, some of them bad."[8] Gregory, like Augustine but two centuries later, was a bishop of the Catholic church and a proud defender of orthodoxy. He challenged not only Jewish visitors who questioned the divinity of Christ but even his own king, whose work on the Trinity contained heterodox views of that difficult doctrine. Gregory also came from an ancient Gallo-Roman family, proud of its ties with the classical world and disdainful of the Frankish conquerors of Gaul. He was sufficiently aware of the grandeur of his classical heritage to apologize for his own lack of literary skill and learning. Neither his Christian orthodoxy nor his classical background seems to have helped him assess the meaning of history. Certainly little of Augustine's complex analysis of time found expression there. Absent from Gregory's work are the twofold direction and underlying unity based on the personal creativity of God. Absent as well is any sense of time as a product of the personal judgment of the individual in creating meaning out of the events.

Though he did not see a specific overall pattern in history, Gregory was still convinced of the importance of temporal sequence. He concluded his preface, "So that the sequence of time may be properly understood, I have decided to begin my first book with the foundation of the world." In fact Gregory attributed more importance to the numerical aspect of this sequence than Augustine himself had. At the beginning of the first book Gregory observed, "For the sake of those who are losing hope as they see the end of the world coming nearer and nearer, I also think it desirable that, from material assembled from the chronicles and histories of earlier writers, I should explain clearly how many years have passed since the world began" (1:Preface). Gregory did not say exactly how his readers would gain hope by knowing the number of years, but he was doubtless considering the common notion that the world would last six millennia to correspond to the six ages. And he did leave his readers with the hope that the end would not come in their lifetimes, since he concluded in the 5,792nd year from creation. Whatever the number meant to him, he was convinced enough of its importance to devote some care to its calculation and to keep his reader abreast of it as the narrative continued.

As guides to his calculation he invoked the authority of three important chronological models. "The chronicles of Eusebius, Bishop of Caesarea, and of the priest Jerome, explain clearly how the age of this world is computed, and set out in systematic form the entire sequence of the years. Orosius, too, who looked very diligently into these matters, made a list of all the years from the beginning of the world until his own day. Victorius did the same thing, when he was making inquiries about the dating of the Easter festival" (1:Preface).

These three models, Eusebius-Jerome, Orosius, and Victorius, used three sequences so different in organization, starting point, and degree of synchronization as to make it virtually impossible for Gregory to create a single sequence out of the concrete events of his story. Even if he had actually wished to implement his theoretical interest in a quantitative chronology, he would have encountered nearly insuperable difficulties. But each of these systems played an important role in the development of medieval chronology. Eusebius organized his *Chronicles* geographically, making only sporadic use of his own chronological tables. He often ended his description of each country with a list of kings, synchronizing one of the kings with the Olympiad or with the reign of the king in another country already discussed. But no single linear series was present in the chronicle. Although Eusebius did calculate the age of the world, he did not use this number systematically to locate and organize events.

It was in the second of the sources Gregory referred to, Paulus Orosius' *Seven Books of History against the Pagans*, that he could find the events of the world in a single temporal sequence organized around a fixed point. A younger contemporary of Augustine, Orosius wrote to show more fully than Augustine that the past was as wretched as the present, and to that end he recorded the natural and civil disasters that had befallen the human race. The resulting catalogue of plagues, wars, famines, and meteorological horrors convinced Orosius "that the days of the past were not only as oppressive as those of the present but that they were the more terribly wretched the further they were removed from the consolation of true religion."[9]

Even though Orosius explicitly wrote to prove the theory set forth in the *City of God*, he did not adopt Augustine's personal sense of time. Instead of the internal process whereby memory and expectation are unified in the present consciousness of the soul, Orosius saw an external process, one in which disasters were constant throughout time. That external process did figure in Augustine's work, and the theme of unremitting disasters even dominated the third book of the *City of God*, but in the work as a whole it took second place to the creative and psychological aspects of time. Orosius severed the external aspect from the psychological one and treated it independently.

Orosius was forging a new path, and he knew it. "Nearly all writers of history . . . have commenced their histories with Ninus. . . . For my part, however, I have determined to date the beginning of man's misery from the beginning of his sin, touching only a few points and these briefly. From Adam, the first man, to Ninus, whom they call 'The Great' and in whose time Abraham was born, 3,184 years elapsed, a period that all historians have either disregarded or have not known" (1.1). Orosius could not claim originality simply by looking at the world from the beginning, for Eusebius before him had started with the creation. The originality lay not in the scope but in the organization. Orosius ordered the history of the world into a single series of events, using a single point of reference so that the reader could see clearly the temporal distance among events. This perspective demanded completeness; all periods of human history needed to be filled in with events.

As a reference point Orosius chose neither the creation, the first event in the Bible and in his own history, nor the reign of Ninus, the first event in Pompeius Trogus' history of empire. He chose instead the founding of Rome. "I shall therefore speak of the period from the creation of the world to the founding of the City, and then of the period extending to the principate of Caesar and the birth of Christ, from which time dominion over the world has remained in the hands of the City down to the present day" (1.1). He located the founding of Rome with reference to known dates of the Hellenistic dating system—414 years after the fall of Troy and in the sixth Olympiad. Events in the first book, covering the period from the creation of the world to the founding of the city are dated before Rome. (Although he claimed to start with the creation, Orosius did not give a fixed date until the reign of Ninus, which he said occurred 1,300 years before the foundation of Rome.) From the founding of Rome Orosius continued his narrative through the early years to the imperial period and down to the 1,168th year of the city, in the waning days of the empire, when Constantius drove the Goths into Spain.

Orosius' contribution can best be seen in a brief comparison with his mentor, Augustine. In both the *Seven Books of History against the Pagans* and the *City of God* the early events of human history have the same conceptual relationship to Rome. Rome became the successor to Babylon, as Christ succeeded Abraham. Augustine, though, expressed this theoretical structure by showing generational relationships among the parts, not by counting the years. Unsatisfied with such vague correlations, Orosius included precise quantitative measures of time. Babylon and Rome were both restored sixty-four years after their histories began (2.2); Abraham was born in the forty-second year of Ninus' reign just as Christ was born in the forty-second year of Augustus' (7.2). Orosius, by quantifying the temporal relationships among the major events of human history, gave narrative content to Augustine's theory and offered a model future historians could use to organize longer and more detailed accounts of the events of the sixth age.

Orosius' interests were practical. He was not content simply to calculate the age of the world and correlate the major theoretical periods. He used the year of the city to locate events not clearly dated in his sources. At one point he cited Tacitus as an authority for his dating of the destruction of Sodom and Gomorrha. "Tacitus, too, among others, mentions that one thousand one hundred and sixty years before the founding of the City the region which bordered on Arabia . . . was burnt . . . by a fire from heaven" (1.5). Orosius was mistaken. In Tacitus' *Histories* there is no explicit date, and since that work is not organized as an annal the date cannot even be extrapolated (*Histories*, 5.7). On another occasion Orosius claimed that Pompeius Trogus dated the famine in Egypt 1,008 years before the founding of Rome. Here Orosius, ignoring Trogus' indifference to absolute dates of any kind, placed the event on his own time line despite his sources' failure to use the founding of Rome as a reference point.

The founding of Rome was useful in creating a single series of events in quantitative temporal relationship with one another, but it suffered from a major drawback. The fundamental event which divided the history of the world in Augustine's schema was not the founding of Rome but the incarnation. Orosius never fully integrated the conceptual structure of Augustinian history with his dating system. Concluding his work, he referred to the coming of Christ as the decisive event in world history. "I have set forth with the help of Christ and according to your bidding, most blessed father Augustine, the passions and the punishments of sinful men, the tribulations of the world, and the judgments of God, from the Creation to the present day, a period of five thousand six hundred and eighteen years, as briefly and as simply as I could, but separating Christian times from the former confusion of unbelief because of the more present grace of Christ" (7.43). When he came to narrate the birth of Christ, however, he dated it simply in the 752nd year of the city. Since the dating system gave no special importance to this event, Orosius felt obliged to include at that point an extended theoretical discussion comparing the temporal relationship between Christ and Augustus with that between Abraham and Ninus (7.2–3). He created a new dating system, one that could locate all human events on a single time line but could not organize them around the major turning point in history.

The Easter calendar of Victorius of Aquitaine, the third dating system used by Gregory of Tours, did not have this defect. It originated, in fact, not as a historical tool at all but as a device to serve the liturgical and administrative needs of the church, and it depended directly on the central event of human history, the life of Christ. Despite the nonhistorical origin of this third chronology, it was to prove more useful as a dating system than either the Hellenistic tables of Eusebius or Orosius' year of the city.

The date of Easter, determined by the first full moon after the vernal equinox, cannot be confidently computed without integrating the lunar and solar calendars.[10] In the effort to accomplish this feat of integration, liturgical scholars during the early years of the church faced a number of problems. Some were purely computational. The numerical relationship between the two cycles is a complex one. Every nineteen years the annual solar cycle coincides with the lunar cycle; that is, the new moon will fall on the same day of the solar year that it did nineteen years ago. But early in the history of the church most of Christendom had agreed that Easter should always be celebrated on a Sunday. Because of leap year, it takes twenty-eight years before the solar cycle is completed and the year is repeated with each month beginning on the same day of the week it did twenty-eight years before. The lunar month and solar year thus coincide with the days of the week only once every 532 years, or twenty-eight nineteen-year cycles.

There were political and administrative problems that were as complex as the computational ones. The Christian church in the early Middle Ages sought to establish a unity even broader than the Hellenistic Oecumene, encompassing different cultures which used a variety of calendars and administrative conveniences. Easter rapidly became a symbol of this unity, for to celebrate the central holiday of the church on the same day was a dramatic indication that the worshipers acknowledged a common faith and religious discipline. As early as the Council of Nicaea, shortly after the conversion of Constantine, the church had decreed that Easter should be celebrated at the same time by all Christians. In order to enforce this

unity the church had to overcome the peculiarities of dating which characterized Europe and the Near East in those centuries and which had survived both the coming and decline of the Oecumene. To this diversity was
added the growing tension between the eastern and western branches of
the church. As part of this rivalry the Bishop of Rome was determined to
establish himself as the head of western Christendom, despite the political
fragmentation which ensued on the disappearance of the western emperors.

To strengthen the unity and to accentuate its own leadership, the Ro-
man church established certain limits in the solar calendar beyond which
Easter could not be celebrated. The church was especially determined that
Easter not fall after April 21, the traditional date for the foundation of
Rome. The celebrations connected with this event tended to be raucous
with pagan overtones, and the bishops found them inappropriate during
Lent. In 455 Easter would have fallen too late in the year by the Roman
calendar, and Pope Leo I moved it to an earlier date. In response to this
crisis a new Easter table was drawn up. Earlier ones had listed the dates of
Easter for a few nineteen year cycles, but in 457 Victorius of Aquitaine
decided to establish a table with twenty-eight such cycles, beginning with
the death of Christ and continuing for 532 years. Apparently he did not
realize that the mathematics made his cycle a repeating one. He adopted it
simply because, as he started from Christ's passion in 28 A.D. and calculated through his own time into the future, he found that the cycle eventually repeated itself.[11] Victorius' calendar defended the April 21 limit by
integrating past Easter celebrations into the present ones and showing the
validity and continuity of the Roman convention.

This table, and others like it, came into widespread use in the next century. They served not only as liturgical guides but also as historical records,
for their keepers began to record on them the events of the year. This use
seems natural to us and hardly a significant departure from earlier practices, but the Easter tables were not like lists of officeholders or priests
which often served as historical records in the classical world. The Easter
tables were primarily a numerical series, lacking the intrinsic relationship
to events that a consular list might have. Unlike the lists, the tables included events not directly associated with their principal subject matter,
the calculation of the liturgical calendar.

Historians of the classical world were in fact not so sure of the relevance
of religious computations to the practice of history. Tacitus was proud of
his own work in computing the period of the secular games which celebrated the founding of Rome. He carefully noted, nevertheless, that he
helped calculate these games in his function as a priest, not as a student of

history.[12] He was disdainful of historians who sought to find significance in elaborate calculations of numerical relationships among events (*Annals*, 15.44).

For medieval thinkers those elaborate calculations expressed a vital aspect of time. The Easter table, by prophesying the occurrence of Easter, located points in the future. At the same time the tables illustrated the passage of the future into the past, for the tables as they filled up with present events became records of the past. Gregory of Tours' own comments about learning the sequence of past years to acquire hope about the future make clear his awareness that the future should be connected to the past. And Gregory was less aware of the psychological aspects of Augustinian time than later historians. As the West acquired a deeper understanding of these aspects, the Easter tables seemed more and more fitting to hold a record of past events. Modern historians describe the process in which the annals grew out of the tables as a natural one, but it depended on assumptions about time that were absent from classical writing.

Victorius' Easter table could not easily serve the purpose of a historical record. Its length and open-endedness argued for it, but its starting point raised problems. Since the table began with Christ's passion, the year had to begin with Easter, a movable date that could only be identified from the table itself. Moreover, the March or April date of Easter did not coincide with the major starting points of years in the ancient world—September for the Greeks and January 1 for the Romans. Gregory of Tours used Victorius' table as a guide for when to celebrate Easter but not as a means of placing the events of Frankish history on a time line (10.23).

During the century after Gregory a fortuitous series of events brought about a table more useful to historians. First, in 525 a Roman scholar named Dionysius Exiguus drew up an Easter table based on Christ's birth.[13] Since this was celebrated on December 25, it fit well with the January beginning of the Roman year. Dionysius abandoned the Greek chronology, based on the accession of Diocletian, because he considered that persecutor of Christians an inappropriate person on whom to base an Easter calendar. In drawing up his calendar, Dionysius knew neither about the mathematical cycle of 532 years nor about Victorius' Easter table. For a model he chose an Alexandrian table which covered the ninety-five year period from 437 to 531 (five nineteen-year cycles). His own cycle of ninety-five years from the end of the Alexandrian table thus began in 532. In 559 Dionysius' ninety-five year cycle overlapped Victorius' 532 year one. At that point the two tables meshed in such a way that the incarnation became the year 1 and the cycle based on Christ's birth became an endless one recording the dates of Easter for an unlimited time forwards and backwards. (The longer cycle itself is an abstraction which does not consider the complexities of the actual lunar and solar years. For the dates of Easter to repeat themselves in the same order in the present calendar requires 5,750,000 years.) [14]

Widespread use of Dionysius' system depended on another Easter crisis, this time in the British Isles, where the Celtic and Roman churches observed Easter according to different calculations. In 665, if nothing were done to prevent it, the two churches would observe Easter at widely separated dates. To resolve this crisis, representatives of both churches met at Whitby in 664 and reconciled the differences. The result was a victory for the Roman church, employing the Easter table of Dionysius, which had lain nearly unused in Rome for a century and a half but which was brought to England to help the cause of Rome there. The English adoption of an endless Easter table based on the incarnation was decisive. During the next two centuries Anglo-Saxon scholars brought that dating system to the Continent, where the calendars were used in liturgical calculations. In addition, lists of major events were added to the calendars, which gradually developed into separate annals. During this period a uniform chronology in the whole world of Western Christendom took hold not only in scholars' annals but even in the official and private documents of everyday life.

This linear chronology based on the incarnation did not preserve all of the personal elements of time which went into its creation. By the period when it was fully formed and supported with sophisticated conceptual defenses, it had become a mathematical series dependent on astronomical observation. Bede, the eighth-century English churchman who constructed an influential chronicle based on the system and who wrote the most elaborate defense of the method of calculating the Easter celebration, looked at time differently than Augustine. Where Augustine was dissatisfied with the astronomical aspects of time, Bede embraced the evidence of the stars as an unshakable proof of the accuracy of his system. In *De temporibus*, he stressed the integration of human and scientific measurement, explaining all of the temporal measurements without reference to the personal consciousness that dominated Augustine's thinking. "Times are divided into moments, hours, days, months, years, centuries, and ages. . . . And in order to avoid error it must be noted that the calculation depends partly on nature, partly on authority, and partly on custom. By nature the common year has twelve months; by custom the month has thirty days; and by authority, the week has seven days." [15]

The Time of the Incarnation

THE MEDIEVAL NARRATIVE

The Year of Our Lord, Anno Domini, focused the measurement of time on the central event of human history but did not stress either the personal nature of time or the individual's experience of the temporal process. Bede's theoretical approach to time was as mathematical as that of Eratosthenes. From a Christian standpoint, such a chronology could organize human history only by leaving out vital aspects of the substance of time and of God's creativity. But historians did not use the Year of Our Lord to construct a single time line. They blended it with Orosius' linear system to create a more complex structure, one which could incorporate episodic narratives as well as a variety of linear perspectives. Their narrative histories encompassed all of human history, but they conveyed the events in patterns which highlighted their inner relationships rather than their position on an external time line. The temporal connection between events was one in which quantitative dimensions were not the only, or even the most important, aspect.

Modern scholars have often failed to understand the complexity of medieval chronology, and that failure has affected their use of sources from the period. As part of the rise of critical scholarship in the nineteenth century, historians turned their attention to the historical writing of the Middle Ages. The resulting studies questioned the accuracy of the medieval texts and refined their usefulness as primary sources. Critical editions and monographs appeared on Orosius, Gregory of Tours, and Otto of Freising, a twelfth-century historian whose *Chronicle of the Two Cities* represents the major medieval attempt to incorporate Augustine's model into a narrative history.[16]

The authors of these monographs and critical editions labored principally to identify those parts of the histories which depended on reliable sources and which thus constituted primary facts that modern scholarship could use in its own reconstruction of the past. Imbued as these scholars were with positivistic assumptions, they were invariably drawn to consider the dating systems of the medieval historians and to complain about their chronological inadequacies. They pointed to inconsistencies, failure to compute figures accurately, and vagueness at key points. Mörner, for instance, regretted that Orosius borrowed dates indiscriminately from his sources, that he lacked an accurate and consistent chronological system, and that he made frequent errors in calculating precise dates or intervals between events (pp. 67–83). Huber claimed that Otto was so imprecise that he must have been indifferent to dates (p. 94).

In these judgments nineteenth-century scholars imposed on their medieval subjects their own assumptions about absolute time and the necessity of a unified dating system. They could not imagine that a historian might prefer to avoid a single time line. Consequently they assumed that some sort of problem with arithmetic or source material must lie behind the apparent inaccuracies they discerned. Indeed medieval historians labored under several such impediments. First of all, before the advent of printing, uniform copies of any work were unavailable; all scholars worked with manuscript editions which varied markedly from the original source. Moreover, the lack of elementary skill in calculation among educated people before the seventeenth century is striking. Tallement in the sixteenth century tells the story of a man who was offered a position as municipal councilor just because he could multiply five times one hundred rapidly in his head. To do computations in Roman numerals is significantly more difficult than it is to do them in Arabic numerals. Not only do errors creep into calculations, but the Roman numerals inhibit the routine checking of results that might prevent mistakes from recurring in subsequent editions.[17]

But the peculiarities of medieval narratives cannot all be traced to these obstacles. Some arise from these historians' commitment to a relative dating system. As the presumed discrepancies are examined in more detail, it becomes clearer that computational errors cannot account for them all. Gregory of Tours, for instance, at the end of his history calculated the number of years that had passed since the creation of the world (10.31):

> From the Beginning until the Deluge, 2,242 years.
> From the Deluge until the Passage of the Red Sea by the children of Israel, 1,404 years.
> From the passage of the Red Sea until the Resurrection of our Lord, 1,808 years.
> From the Resurrection of our Lord until the death of Saint Martin, 412 years.
> From the Death of Saint Martin until the year mentioned above, that is the twenty-first after my own consecration, which is the fifth year of Gregory, Pope of Rome, the thirty-third of King Guntram and the nineteenth of Childebert II, 197 Years.
> This makes a grand total of 5,792 years.

This calculation is not without importance. The final figure of 5,792 is a comforting one which gave the hope Gregory promised at the beginning of the work. It comes short of the 6,000 years which many thought would signal the end of the world—a millennium for each age—but it is not the sum of the the figures Gregory gave. The individual figures in the passage add up to 6,063, a much more disturbing figure. If 6063 is correct, then either some of the figures themselves are a little too large, making the total just under 6,000 and the end of the world in immediate sight, or the figures are correct and the theory is wrong. In that case the meaning of the whole chronological series Gregory used comes into question.

How did this error insinuate itself into Gregory's writing? The critical scholars, feeling that Gregory himself could not have made such an error in the first place, suggest the copyists.[18] But the same objections can be raised for copyists' errors. If such an error remained, then neither the copyist nor his subsequent readers could have thought the actual linear series of great importance. What was important to them was the final figure. Its relationship to the series of events that made up the narrative was only theoretical, having no direct part in the actual business of constructing a narrative.

Similar attitudes towards computation can be found in other historians. Otto of Freising, for instance, gave alternative numbers for the intervals from Adam to Abraham, citing both Jerome's figures and those of the Hebrew Septuagint.[19] Like Augustine before him, Otto felt the precise number was not important (City of God, 15.20). What concerned him was what happened in that period and how its meaning could best be understood and expressed. To penetrate that meaning he did not need to know an exact number; it sufficed to know that such a number existed and that it was part of a linear series of world events from Adam to Otto's own day.

If arithmetical errors cannot explain these peculiarities, neither can problems with the sources. Though medieval historians worked with several sources and were faced with a variety of chronologies, they approached the problem with imagination and a consistent goal. They combined diverse sources to create a coherent narrative whose temporal dimensions were appropriate to their theme. In the first two books of Orosius' history for example, the historian faced the problem of integrating several sources with different chronologies. He concentrated in these two books on the argument that both Babylon and Rome, the two great empires of human history, had suffered many disasters. To express this theme he divided the two books chronologically, treating the history of the world from the creation to the founding of Rome in the first book and the history of Rome from its founding to the sack of the city by the Gauls in the second.

The two books rely on different sources. In the first, Orosius drew his stories mainly from Justin's summary of Pompeius Trogus, while in the second he used Florus' epitome of Roman history, with occasional Roman events taken from another epitome by the fourth-century scholar Eutropius. He continued in the second book to rely on Justin for events in the East. Since neither of his principal sources used absolute dates systematically, Orosius needed help in tying the two books together, which he sought in the chronicle and chronological tables of Eusebius as translated and extended by Jerome.

In converting Eusebius' variety of dates into his single time line extending forward and backward from the foundation of Rome, Orosius made a number of egregious chronological errors, particularly when he brought into the narrative events from subsidiary empires such as Persia and the various city-states of Greece. Following Justin's narrative, he described several Greek wars, including the Spartan conquest of Messena, as if they had happened before the foundation of Rome. Not only that, but he told the story of Cyrus in both books, dating him the first time just before the founding of Rome and the second time 200 years after it. Finally, he narrated the Peloponnesian War twice, locating it in the first book just before the founding, and in the second 350 years after it.

The fact that he relied on several sources does not fully explain these repetitions and errors, for Orosius used Justin throughout both books for his account of non-Roman events. The discrepancies do arise partly from his attempt to reconcile the narrative in Justin with Eusebius' tables, but even that problem does not suffice to explain the order of Orosius' narrative. Eusebius listed most of the events, even though he did not cover the Peloponnesian War clearly in his own chronicle. Eusebius' tables plainly date the Messenian war in the tenth year after the founding of the city, not twenty years before, where Orosius placed it.

A superficial reading of the text might lead one to conclude that Orosius was led astray by the traditional division of the narrative between East and West. He certainly was concerned with the geographical and spatial aspects of his narrative, beginning his work with a description of the world that became the standard source among succeeding generations of historians (1.2). Spatial considerations influenced his introduction of the Peloponnesian War in both books. In Book One he talked of the war from the Greek point of view, as Sparta attacked Messena, while in Book Two he mentioned Sicilian affairs in context of the growing Roman power. But this organization breaks down on closer inspection, since Sicilian affairs in Book One were introduced before he talked of Greece, and in Book Two Orosius devoted a long section to the events in Greece that followed the

Sicilian campaign (2.16-17). The Eastern and Western perspectives are thus present in both books.

Orosius organized the temporal aspects of his narrative thematically, not spatially or quantitatively. He measured time by theme rather than by number. His practice is clearest in the aforementioned repetition of the story of Cyrus. In Book One Orosius introduced him as part of the rise of the Empire in the East. Cyrus was shown there overthrowing his predecessor and conquering the Medes, thus completing the final stage of the Empire in the East, its translation to the Persians, just as Rome was coming into existence in the West (1.19). In Book Two, Cyrus was introduced not to show the linear time of political change but to illustrate that the calamities of the pagan world were continuous, even while the Empire in the East was falling and that in the West was rising. In Book Two, therefore, we see that Cyrus ravaged the whole East at the same time Tarquin the Proud was oppressing Rome and that he took Babylon while the Romans were losing one of their most important families in a war with their neighbors (2.6). In Book Two Cyrus acted in episodic time. His acts contributed to a theme whose temporal significance depended on duration rather than a linear series of points. Both themes—translation of empire and constancy of disasters—have temporal dimensions and both are plotted on the same dating system. But Orosius conceived them as separate, overlapping themes, one existing in linear time and one in episodic time. A particular event relating to both themes need not have a single place on the time line.

Not all such discrepancies in Orosius' work concern events from the distant past. In dating specific periods from the reigns of recent emperors he did not always give the same length to an emperor's reign. In one passage, for instance, it appears that Vespasian reigned only three years, from the 825th to the 828th year from the founding of Rome. In the narrative of that emperor's reign, however, Orosius described an earthquake as occurring in the ninth year (7.9). The two figures occur too closely together in the text to be the result of simple carelessness. Nor can they be easily traced to a discrepancy in his sources. He found the earthquake in Jerome's extension of Eusebius' chronicle, but he could not have taken his dates for Vespasian's reign from that source. Even though Jerome did not give a year-by-year list of the events from the founding of Rome, he made clear that Vespasian reigned at least ten years. Nor was Orosius relying on Eutropius, who did not date Vespasian's reign at all and whom Orosius corrected on other occasions.[20] Orosius simply did not conceive of the two figures as part of the same time sequence. Like Cyrus, Vespasian acted in two times, one of empire, where a numerical series existed, and one of disasters, which could be dated as episodes in a specific emperor's reign. One of these times was linear and one was not; the connection between them was not significant.

Orosius' sense of time, despite the persistence of episodic and linear qualities, differed from that of the Hellenistic historians. For Tacitus and Livy, events in the West and East existed in separate times, with participants moving from one time to another as they traveled in space. For Orosius, the different temporal sequences were not spatially divided. The *translatio imperii* moved from East to West as part of a single linear sequence. But in grouping the different regions into a single time, Orosius did not create an absolute time line. Beside this linear time of empire there was another time in which the catastrophes so important to his defense of Christianity were constant. The numerical errors in his account arose from his attempt to group events thematically in their true temporal relationship. The dating system only forced into the open a faulty synchronization that was implicit in Pompeius' work but was not easily seen in his chronology.

Later medieval historians, using the incarnation rather than the founding of the city for a dating system, also felt that the quantitative aspect of temporal relations was less important than their thematic unity. They doubted that this thematic unity could be adequately conveyed by locating events on a single time line. Consequently they used a variety of other dating systems to bring out the particular temporal meaning of events. Even Bede, who formulated the theoretical defense of the chronology based on the incarnation and drew up chronological tables showing how his system could encompass the entirety of human history, used several different chronologies. He allowed the incarnation to stand by itself only for ecclesiastical events, such as the death of an archbishop, or astronomical events, such as an eclipse or a comet.[21] The astronomical and ecclesiastical issues surrounding the Easter calendar would have been vividly present in Bede's mind as he rehearsed the history of the church, since it achieved its unity by resolving these particular issues. Thus the incarnation expressed the essence of the meaning contained in such events and could serve as the single means of temporal location.

For events associated with the rise and fall of Roman power in England, Bede used several different references. He dated the conquest of Britain from the foundation of Rome (1.3) and the sack of Rome from both its foundation and the conquest of Britain (1.11). By that means he numbered a temporal series of the rise and fall of an empire, a theme taken from such earlier historians as Orosius. Bede dated the fall of the Roman

Empire in the West by reference to the reigning Eastern emperor, thus creating in the reader's mind a second political dating system, with connotations of ongoing power. In addition, he kept track of the reigns of Roman emperors, locating events in the British Isles according to those reigns.

For events he felt were unusually important Bede synchronized these different sequences according to their meaning. He dated Pope Gregory, who initiated the conversion of England but who was also a major figure in Rome itself, both by the year of the incarnation and by the reigns of the Roman emperors (2.1). The Synod of Whitby, with its political, cultural, and religious significance, is dated by the incarnation, the reign of King Oswy, and the accession dates of the bishops (3.26). The final Roman conquest of England under Claudius, permanently tying the island with the Roman world, possessed special significance for Bede. He dated it not only from the founding of Rome, the reign of the emperor, and the incarnation, but also referred to a famine mentioned in the Book of Acts as happening at the same time. By mentioning that event, which he found correlated with Claudius' reign in Eusebius, Bede integrated the conquest further into events of the Roman world (1.3).

The year of the incarnation played for Bede much the same role that the Olympiad did for Polybius; both were limited in application to a coherent and fundamental theme. He saw the system as a useful tool for ecclesiastical administration and as a means of reinforcing the religious unity of the British Isles. He was certainly capable of using it to create a linear series of events, and at the end of his *Ecclesiastical History*, he included such a list, dated from the incarnation. But particular events still needed to be located in relation to one another by a chronological reference which would bring out their specific thematic relationship. To identify the year, for Bede, was not simply an act of numerical calculation. Dating involved judgments of moral worth and spiritual significance. After describing the vicious reign of King Eanfrid, Bede observed, "This year remains accursed and hateful to all good men, not only on account of the apostasy of the English kings . . . but also because of the savage tyranny of the British king. Hence all those calculating the reigns of kings have agreed to expunge the memory of these apostate kings and to assign this year to the reign of their successor king Oswald, a man beloved of God."[22]

Bede used several dating systems partly because of the novelty of the chronology based on the incarnation. His readers would be unfamiliar with it and would not be accustomed to using it in their own documents. This unfamiliarity does not fully explain his practice, however, since he used the other dating systems in a consistent and systematic way, stressing the intrinsic temporal relations among the events being dated. In the centuries after Bede, the year of the incarnation became commonplace as a means of dating the documents of everyday life as well as of the annals recording events from year to year. By the twelfth and thirteenth centuries readers were thoroughly familiar with it, but many historians still supplemented it with other dating systems.

Some writers used other references in ways that show insecurity about how precisely the year of the incarnation can fix an important event in time. The thirteenth-century crusade chronicler Robert of Clari, for instance, began his history of the Fourth Crusade, "Here begins the history of those who conquered Constantinople. . . . It happened, in that time when Pope Innocent was apostolic of Rome and Philip was king of France and there was another Philip who was emperor of Germany and the year of the incarnation was one thousand two hundred and three or four."[23] But more sophisticated historians than Robert, writers who were more sure of their dates, also used the incarnation in conjunction with other temporal references in order to bring out thematic relationships. Otto of Freising used three dating systems. He dated the events up to the founding of Rome from the reign of Ninus, and used the foundation of Rome to measure subsequent events until the birth of Christ. From the birth of Christ he used the incarnation or the founding of Rome depending on the meaning of the event.

Otto saw a linear connection between these three dating systems. Like Orosius and Augustine, he presented the fall of Babylon as a prelude to the rise of Rome (2, Prologue), and he was convinced that the unity created by the Romans prepared the way for Christ's birth (3, Prologue). He chose the three dating systems in order to convey the essential meaning of the events within each sequence and to separate the three clearly from one another. "Let us begin from Ninus the story of human misery, and let us cry it through, year by year, from Ninus himself down to the founding of the City; from the founding of the City let us relate it in due order as far as the time of Christ and from the time of Christ, by God's help, likewise down to our own day."[24] It is this sharp division that determined his choice of dating systems. As he said in adopting the incarnation as the start of the third sequence, "So then, with the birth of the new Man who supplanted the old, let us set an end to our annals also, which were reckoned from Ninus to the founding of the City and thence even to this time, and let us begin our annals from His birth" (3.6).

Otto was not simply using three chronologies to the same purpose that we use a single one. The three sequences, though more clearly delineated than in the works of Orosius and Bede, overlapped much as theirs had, and Otto did not join them into a single linear system. He continued to use the Roman sequence after the birth of Christ to record the subsequent history of the Roman Empire, noting like Orosius that Alaric sacked Rome in the 1,164th year from the founding of the city. Where Orosius had dated the sack only in the Roman sequence, Otto was more interested in showing the relationship among the three sequences. He mentioned the year of the incarnation in which the sack occurred and remarked that the same period of time had elapsed between the founding of Rome and the sack as had elapsed between the reign of Ninus and the destruction of Babylon (4.21).

Synchronization was important to Otto but the time line alone could not express its significance. Simultaneous events needed to be linked thematically and concretely, not simply by location on one of his linear sequences. In Book Two he wondered whether Carthage was destroyed at the same time that Simon Maccabeus was leading the Hebrews. He could calculate from his sources how long after the founding of Rome each of these events occurred and thus express their relationship through an absolute date, but the connection between the two involved moral and spiritual issues. It was fitting that "the Maccabees, those notable zealots in behalf of the law of God, and the Scipios, the renowned defenders of the city of Rome and of the laws of their native land, are found to have been contemporaries" (2.42). Since the simultaneity was thematic, however, it needed an actual physical connection, not simply a theoretical one established by the dating system. To show that this appropriate relationship existed in fact Otto described a shield which was sent by Simon to the Roman Consul Lucius in the 172nd year of the rule of the Greeks in Palestine, dating both the destruction of Carthage and the priesthood of Simon by the Greek calendar. By this means he had proved the synchronism between the destruction of Carthage and the revolt of the Maccabees, but he needed now to place this synchronism on his own time line, which he accomplished in a long and complicated passage comparing various dates in the Greek world with the Roman equivalents. He established the Roman dates in order "that what I have said may become more evident [apertiora]" (2.42). Dating for Otto clearly involved complex thematic and moral issues, not simply computational ones.

Otto's narrative is more linear than those of the earlier historians. The three major dating systems overlapped, but within each sequence he confidently established patterns and synchronisms across geographical and political lines. Where Orosius had seen the same emperors participating in different times whose quantitative dimensions were incompatible, Otto compared lengths of reigns to create a linear sequence to support his historical judgments. He compared sources, for instance, to establish the temporal relationship between the reigns of Cambyses and Tarquin (2.15).

In later parts of *The Two Cities* Otto described events closer to his own time, events where the participants themselves were conscious of the Christian dating system. Even there he did not use the Year of Our Lord to create a single linear time. Instead he created distinct political and spiritual sequences, both based on the incarnation, while using more traditional generational systems to locate events in nonlinear episodic time. The political sequence in the last books of *The Two Cities* records the series of Roman emperors in a continuous line from Augustus. To create that sequence he measured the coronations or elections of the emperors from the incarnation in precise numerical terms. "In the one thousand one hundred and thirty-eighth year from the incarnation of the Lord, since the emperor Lothar died in the autumn without sons, a general convention of nobles was appointed to meet at Mainz on the following Whitsunday" (7.22; see also 8, 11, and 17). These specific references to new emperors account for nearly all the dates in Book Seven.

The spiritual sequence in Book Seven is less precisely dated than the political one and has less connection with concrete events. For religious events Otto created a time series marked by round numbers from the incarnation, thereby conceiving it as if it were an abstract reality significant in its own right independent of the concrete events that illustrate the passage of spiritual time. "In the one thousand and one hundredth year from the incarnation of the Lord, when the faithful were flocking from all parts of the world to the earthly Jerusalem (the counterpart of the heavenly) to pray, many died from the unwholesome climate" (7.7; see also 7.30). The events in this religious sequence serve to illustrate its theological meaning and thus are chosen according to an arbitrary year, just as Eusebius had indicated years from Abraham by ten-year intervals or Gregory of Tours had calculated the total number of years from the creation. Ecclesiastical issues are often treated in this sequence, even if they have clear political significance. To describe the Investiture Controversy Otto grouped events from all over Europe into units of a decade. Otto's conception of the Year of Our Lord in referring to events of spiritual significance contrasts clearly with his usage in dating political events. In the political sequence the events had a meaning themselves and were consequently recorded according to their importance in the year in which they occurred. In the spiritual sequence the time itself is the locus of meaning, and events are chosen to illustrate its passage toward the fulfillment of the spiritual destiny of humanity.

The existence of separate spiritual and political sequences in the last part of *The Two Cities* reflects one of Otto's most important themes, for after Constantine he could no longer talk of two separate cities—one of man and one of God. He felt, however, that though the city was now one, it was composite, grain mixed with chaff (5, Prologue). Thus the relation between spiritual and political domains had become partly symbiotic and partly dialectical. There was a growth in spirituality and wisdom with the rise of the Frankish state and the Holy Roman Empire, but since the church and state could be political rivals weakness in state power often accompanied a strong church. When a bishop refused to let the Emperor Lothar divorce his wife, Otto challenged his readers to "observe now that as the State declined the Church became so powerful that it even judged kings" (6.3). The separate spiritual and temporal sequences helped Otto narrate this complex relationship between church and state.

These two sequences do not include all events. Many themes in the work do not need linear orientation, and for these Otto did not show the location in Christian years. Instead he allowed the events to flow according to their own order, connected with phrases like "at that time," "while such things were happening," or "about this time." When he felt the need of temporal orientation during such a narrative, he often located events within the reign rather than in the linear system of the incarnation. "As [Emperor] Lothar was returning from Italy he was taken sick at Trent, and there in the mountains, in a lowly hut, this mighty emperor died, full of days, in the thirteenth year of his reign and in the seventh of his imperial power, leaving behind him a pitiable memory of our human estate" (7.20).[25]

Episodic, nonlinear time played an important role in Otto's general understanding of the historical process. One of his favorite metaphors for change was the sea, with its unpredictable storms and recurring waves. Even in the major turning points in the history of the city of man, such as the rise of Rome itself, he saw disaster alternating with divine favor. Taking a passage from Orosius (6.14) as his inspiration, he said,

> We see at what cost . . . the Roman Republic grew. By alternating changes after the manner of the sea, which is now uplifted by the increases that replenish it, now lowered by natural loss and waste—the republic of the Romans seemed now exalted to heavens by oppressing nations and kingdoms with war and by subduing them; now in turn was thought to be going down again into the depths when assailed by those nations and kingdoms or overwhelmed by pestilence and sickness, and . . . after they had arranged everything else well . . . they were miserably disemboweled by falling upon one another in internal civil strife. (2.51)

Still later, when he discussed the breakup of the Frankish kingdom: "The world, after the manner of the sea, threatens with destruction by times of storm as the sea does by its waves those who entrust themselves to her" (6, Prologue). In both cases Otto used the metaphor of the sea to encourage his readers to think of realities outside the linear sequence of political events that dominated his work. "All these calamities should have had the power to direct people to the true and abiding life of eternity" (2.51).

The element of alternating fortune and disaster existed in another time. It had a clear temporal reality, and Otto directed much of his analysis to explaining its significance. He pointed directly to the deficiencies of a linear narrative and stressed the importance of presenting the full dimensions of the human condition. "I did not merely give events in their chronological order, but rather wove together in the manner of a tragedy their sadder aspects" (Dedication). At the conversion of Constantine he discussed at some length the relation between the former adversity of the church and its present prosperity, making clear to his readers that the change did not alter the church's status as the City of God. The growth of human wisdom, which was itself part of the linear series, prepared men's minds so that they could experience prosperity without losing sight of the eternal goals of humanity or surrendering their commitment to contemplation as a path to eternal bliss. Despite these linear qualities, Otto considered this section to be a digression from the events he had been recording in linear time and concluded the section on the conversion of Constantine by saying, "now let us return to the sequence of history [ad hystoriae seriem]" (4.4). The issue was not part of the linear series that traced the growth of political power and its relation to the church.

Otto used the metaphor of weaving to describe the historian's work, and he wove a complex texture of time into his narrative.[26] First he used two sequences of dates—from Ninus and from Rome—to describe the linear order of political change and its relation to spiritual matters before the incarnation. Second, he used the incarnation as a starting point for a linear sequence that measured political changes in Christian states. Since the sequence from the founding of the city and that from the incarnation overlapped for several centuries between the birth of Christ and the sack of Rome, he used either or both depending on the meaning and complexity of the event being dated. Third, he included a spiritual sequence beginning with the incarnation and expressing realities which transcended historical events. Here Otto measured time abstractly, using events only to illustrate the contrast between the continuing misery of man and the delights of the heavenly Jerusalem. Finally, he used an episodic time, measured less precisely, to express major themes of the narrative, especially the continuing misery of men. This theme became central to Christian historiography after Orosius, but it had never been absent from the writings of Western historians. From Herodotus to Otto they gave similar temporal dimensions to it.

THE LIMITS OF AUGUSTINIAN TIME

The incarnation, like the Hellenistic Oecumene, produced a new time and a new chronology. Just as the Oecumene had lengthened the chronological scope of the historians' vision, so too did the incarnation. Augustine conceived of time as a product of mental synthesis, beyond the power of measurement, limitless in scope and bounded only by the beginning and ending of human life. This concept of limitless and universal time proved difficult to implement, especially when treating of general history outside the framework of an individual human life. To organize history as a whole Augustine used Age theory, a conceptual structure of little use to a narrative historian.

The actual chronology of medieval historians developed from other roots and embodied other concepts. It reflected the accomplishments and needs of the Oecumene more than the concept of personal time. It represented an enlargement and fulfillment of the Hellenistic search for a comprehensive dating system, one which could measure and integrate many polities into a single oecumenical process. No single system sufficed to measure this new process, and medieval historians gradually settled on two major chronologies. One, dating from the foundation of Rome and called "the year of the city," applied to the growth of empire and measured essentially the same political time as the Oecumene but with a more systematic linear dimension. Orosius recorded the same movement of empire from east to west that Pompeius Trogus had described, but he gave number and quantity to a movement which in the Hellenistic historian's works was geographic and metaphoric. The second chronology, dating from the incarnation and called "the Year of Our Lord," measured spiritual time, including the growth of Christian states. This second system was more consonant with Augustinian theory, and as Christians became more comfortable with the notion of personal time it seemed more natural to record events of the past on a calendar which predicted the future. The idea that future and past were both mental concepts helped liberate historians from feeling that political events had to be recorded on a purely political time line. But that liberation came slowly.

Theory and practice did not always coincide. Just as Eratosthenes' mathematical model did not seem right to practicing historians of the Hellenistic period, so Augustine's universal theory of time did not seem to fit the events. Historians used a variety of dates to record the many sequences they saw in human history. The sequences theoretically embraced all of human history from beginning to end. Since they overlapped, events existed in more than one sequence and had more than one temporal significance. Only in hindsight can we call the Christian year a universal time line. None used it that way in the millennium after Augustine.

New times did not replace the old; they overlaid one another. The episodic time that helped Herodotus organize events to express some of his major themes also gave order to important themes of Christian historians, such as the continual occurrence of human disasters and the moral judgment of God in punishing the wicked. The linear time of Polybius, with its oecumenical scope, plotted the growth of empire and its movement from east to west. The religious time represented by the year of the incarnation was also linear, but it charted a different sequence, the new dispensation in political and ecclesiastical life that the incarnation brought into being.

Historians did not find these dating systems useful to organize the personal life that is the core and substance of Augustinian time. They seldom employed absolute dates of any kind to measure the lives of individuals. Otto's biography of his nephew, Emperor Frederick I, referred only once to an absolute date, when he dated the emperor's accession in the year from the founding of Rome, placing him in the political sequence of translatio imperii, and from the year of the incarnation, placing him in the sequence of Christian rulers.[27] On rare occasions Otto dated events by the year of Frederick's reign, but these were invariably acts connected with his public functions, such as when he was crowned or resolved a political dispute (*Gesta*, 2.32,55). Instead of fixed dates, Otto mentioned other events or rules in relation to which Frederick's deeds occurred, connecting them with vague temporal references such as "at about the same time," "after this," and "while this was going on" (*Gesta*, 2.10,15,33).

Personal time did not fit into the established linear sequences of medieval historians. Their descriptions of persons lacked some essential elements of Augustinian theory, elements that gave such meaning to his autobiography. Individuals in medieval narratives do not interact concretely with their environment, they lack the personal unity that gave coherence to Augustine's development, and they have little direct connection with the general political and spiritual processes that are expressed in linear time. Classical categories of motivation and moral judgment predominate, even though there is a trend toward greater coherence than was present in classical biography. Both Orosius and Otto resolved the contradictions in the classical biographies of Tiberius. They presented a developmental picture in which the Senate's resistance to his attempt to Christianize the empire corrupted the early virtues and created the vicious old man (Orosius, 7.4; Otto, *The Two Cities*, 3.9–11). Einhard's biography of Charlemagne, though modeled on Suetonius' life of Augustus, nevertheless drew a more coherent picture of his subject's character than he found in his classical model.[28] Clearly the linear qualities of Augustinian time made the episodic nature of classical biography seem unnatural, just as the historians of the Oecumene found the episodic quality of Greek political narrative strange.

Medieval historians, despite the greater coherence of their picture of individuals, still tended to isolate them from the political and spiritual time series that dominated their histories. Individuals existed in an episodic time of reversal of fortune and of human disaster, not in the linear time of political growth and spiritual development. This separation is true even among historians such as Otto, who deliberately sought to implement the Augustinian view of history. As part of his account of the breakup of the Frankish kingdom, Otto had to discuss the career of Charles the Fat, who was deposed and beggared after a long and successful reign. Charles existed in the time of *translatio imperii*, where Otto dated his deposition clearly. "In the eight hundred and eighty-seventh year from the incarnation of the Lord, since the emperor Charles had commenced to fail in body and mind, the princes of the kingdom brought to the throne Arnulf" (*The Two Cities*, 6.9). As part of this political sequence, Otto had previously described Charles' strong and decisive policy as emperor (6.8), and he went on in the subsequent section of the narrative to assess how Charles' deposition affected the transfer of power. "And so the kingdom . . . of the Franks lost . . . Belgic Gaul, including Aachen, and the greater part of Francia" (6.18).

Otto interrupted this linear account of political change to describe the significance of Charles' personal fate, which existed in another time and was not directly related to his political successes. His deposition by the princes was not the result of political misjudgment or the accumulated resentments of an unjust reign. His senility and poverty illustrated the reversal of fortune which gave significance to so many events outside the linear time of political change. "Strange to say you might have seen the emperor who after Charles the Great had the fullest power among the kings of the Franks, in a brief space of time [in brevi] reduced to such insignificance that in need even of bread he was pitifully begging a stipend from Arnulf. . . . Behold the wretched condition of our mortal lot" (6.9). Charles was not punished for his sins, for "it is said . . . he was a very earnest

Christian. Wherefore this trial is believed to have come to him at the last to test him," and to show Otto's readers the folly of seeking happiness in the things of this world. Though Charles' deposition was dated in the linear time of political change, his death is measured in the personal time of his own life. "He died in that year, reigning six as king and one as a citizen." Otto showed his theoretical grasp of personality not through his narrative but by digressions into philosophy, especially in his life of Frederick I. This most Augustinian of the medieval historians still felt history to be a lesser discipline and that the highest knowledge was only attainable by abstract philosophy (*Gesta*, Prologue).[29]

The failure to incorporate individuals into the time line is crucial, for the comprehensiveness of the linear series depends on the notion of time as a mental synthesis and on the comparison between our own perception of time and God's. Because this notion did not play a vital role in forming the chronology of medieval historians, their time could more easily blend with the times of the ancient world to form a complex tissue of different times and different dates. Only by personalizing the time line could Western historiography arrive at the sense of a single time, embracing all possible human experience and comprehending all other temporal references. Such was the goal and the achievement of the Renaissance historians.

6

The Time of the Renaissance

> In the upper world on the right bank of the Adige, in the city of
> Verona beyond the Po, on the sixteenth day of the Kalends of July, in
> the 1345th year from the birth of the God whom you did not know.
>
> Petrarch, *Le familiari*, bk. 24, letter 3

PETRARCH AND THE DISCOVERY OF THE ANCIENTS

When the canons of the cathedral at Verona asked their famous visitor to
look over their extensive library, they hoped he would identify some of its
manuscripts. By 1345, when he stopped at Verona, Petrarch already en-
joyed a considerable reputation as a scholar and writer, partly from his edition of Livy's history of Rome, partly from his own writings on Roman
history, and partly from his far-flung correspondence. He came to Verona
on his way back from Rome, where he had been the first poet since ancient
times to be crowned poet laureate. The scholars of his age acclaimed him
as the major defender of those humanist and literary studies that would
dominate the following centuries of Renaissance and revival.

Petrarch did not disappoint his hosts. He discovered in the library at
Verona a major document from the classical world, one which would affect
his own life, alter his concept of humanity, and give him a new sense of
time. He found Cicero's letters to his friend Atticus, written while the
poet was staying outside Rome during the tumultuous time of civil strife
before the coming of Augustus. In these letters Cicero gossiped freely on

all sorts of things, about life at Rome, about his personal affairs, and, most tellingly, about his role in Roman politics.

The Cicero who revealed himself in these letters was not the same Cicero that medieval thinkers had known. That medieval Cicero was a sage and philosopher, the man whose philosophical summaries of Hellenistic thought had converted Augustine to the pursuit of wisdom. Before that day in Verona, Petrarch, like generations of scholars before him, had looked at the writings of Cicero as an abstract and unattainable ideal of self-denial, contempt for things of this world, and devotion to contemplation. As Petrarch read the letters to Atticus, he encountered an active and vindictive politician, who could flout the laws of Rome to execute his enemies and who spoke more passionately of patronage and wealth than of virtue and truth.

Petrarch's first reaction was shock and dismay. In his turmoil he wrote Cicero a letter bemoaning his disappointment. "I searched long and avidly for your letters, and having found them where I least thought to, avidly read them. I heard you, Marcus Tullius, saying much, complaining of much, in various moods. I already knew what a teacher you were for others; now at last I learned that you did not know how to guide yourself." Why, his disillusioned admirer asked, did Cicero not use his own philosophy in his personal life? Why did he leave retirement to enter politics, and why did he abuse such great men as Julius Caesar and Brutus? "How much better it would have been," Petrarch lamented, "if you had never held the fasces of power, sought triumphs, or corrupted your soul with any Catilines."[1] On further reflection Petrarch realized the cause of his disappointment. He had regarded the writings of Cicero as abstract formulations of the truth, unrelated to any specific personality. He had not understood that Cicero was in fact a person as well as a philosopher, that he not only thought but lived and breathed. "If I can speak freely, Cicero, you lived as a man, you spoke as an orator, you wrote as a philosopher. I censured your life [in the previous letter], not your genius, which I admire, nor your tongue, which amazes me. And in your life I ask only constancy, the desire for quiet which is suitable for the practice of philosophy, and that you avoid civil strife" (4:24.4). Petrarch's insight that ideas are not simply abstractions but products of living people in a historical context seems hardly a formidable achievement to modern readers, but earlier writers did not see ideas in that way. They certainly understood that ideas came into existence as part of a temporal process and that they were the product of specific minds. Because of this temporal connection, medieval writers agreed that ideas were a proper subject of historical study. The *Chronological Canons* of Eusebius are full of the names of literary and philosophical figures. Augustine used Eusebius to calculate the temporal relations between Plato and the Old Testament writers. In the *City of God* he refuted the theory that Plato read the Scriptures in Egypt, by showing that he died seventy years before the Scriptures were brought there (8.11). Medieval historians gave beginnings even to specific philosophical techniques. Otto of Freising included important methods of thinking in his history, mentioning that Aristotle had invented the syllogism (*Two Cities*, 2.8).

To medieval thinkers these intellectual achievements occurred in a different time from that of the personal lives of their creators. They occurred not in the paradoxical time of personal growth but in the linear time of increased human wisdom, where God was preparing humanity for the coming of Christ. In this linear time, knowledge, like power, moved from one culture to another. With firm roots in the Eusebian chronicle tradition and tracing its lineage to Herodotus and Thucydides, the idea of intellectual transference found expression among Carolingian writers, and it became commonplace in the Middle Ages.[2]

Though they accepted the transfer of wisdom as an element in the historical process, the writers of the time admitted no direct relation between it and personal time. Augustine could see that his own personal life, like that of his friends, was a dynamic process in which specific writings, like Cicero's *Hortensius*, could change fundamental values and motives. But he could not see that the *Hortensius* itself was the product of the same dynamic interchange between Cicero and the environment of late republican Rome. In the passage from the *City of God* where he showed that Plato could not have read the Scriptures, Augustine went on to say that the source of the ideas in the Platonic writings was not important. The ideas alone mattered to Augustine; their source did not affect their content. For him the question of temporal location was unrelated to the meaning of the text; it only arose in connection with the external issue of the meaning and direction of human history as a whole.

In contrast to the medieval concept of intellectual change, Petrarch's was both personal and intellectual. It brought together into a single time frame entities that medieval historians had kept apart. He sought to express this new temporal reality in the conclusion of his letter to Cicero. "In the upper world on the right bank of the Adige, in the city of Verona beyond the Po, on the sixteenth day of the Kalends of July, in the 1,345th year from the birth of the God whom you did not know."

The use of the dating system here is peculiar. Petrarch was clearly trying to express the connection between himself and the Roman orator. To do so he mentioned places both he and Cicero would be familiar with and identified the time of the year in terms both would recognize. But he did not measure the time between the two in terms that Cicero would have understood. Petrarch could certainly have chosen a dating system familiar to Cicero, had he wished to. The common medieval usage of dating from the founding of Rome would have served well here. Bede used it when he dated Julius Caesar's consulship "693 years after the founding of Rome and 60 years before the birth of our Lord" (1.2), and the practice was well established by the late Middle Ages. But the year of the city expressed a different time from the personal time that connected Petrarch and Cicero. To count from the founding of the city to the date when Petrarch wrote would leave the relationship in the political time of the Oecumene; to count as Bede had done in the two chronologies would divide the relationship into two separate entities. It was precisely these political and spiritual aspects of life, which remained separate in medieval chronology, that Petrarch saw as a single temporal process, and his insight bridged the gulf that divided the two times in traditional historiography.

As he saw the spiritual and cultural achievements of the world in a single time frame, Petrarch found he could make new and startling comparisons between the world of the ancients and his own. In the process he saw what centuries of scholars had not. He saw that Rome as a cultural and political entity had long disappeared from the world, that it had been succeeded by a new age. This was not the sixth age of Augustine and medieval theorists, not an age of light and understanding in enjoyment of God's revelation, but a dark age of barbarism and cultural deprivation (4:20.1;24.8).[3] Petrarch had little interest in this dark age, the period we now call the "Middle Ages," and was not even sure that his own age had emerged from the darkness. Only the glorious times of Rome deserved to be called history; the rest was hardly worth study. "What is all history other than praise of Rome?"[4]

Petrarch's peculiar way of explaining his temporal relationship to Cicero marked a step in measuring the new time. Petrarch saw in Cicero's relationship to Christianity historical dimensions, not simply a theological and conceptual one. The non-Christian status of the ancients became part of their personal history, not simply an intellectual issue marking them off irreconcilably from believers. In subsequent letters to other figures from the ancient world, Petrarch's concern over their personal relationship with the incarnation and the early church became even more explicit. To Livy he said, "In the 1351th year from the birth of Him whom you would have seen or of whose birth you would have heard if you had lived a little longer" (4:24.8).

The fact that these writers were not Christian raised special problems for Petrarch. He certainly wished to respect the thought of the classical world, as the generations of medieval scholars before him had respected it. But he also wished to see the ancients as persons, and for this he had to evaluate their spiritual life and personal attitude toward the values that a Christian society considered crucial. Rejection of these values was not so important when classical writings represented merely expressions of abstract truth, but now that Petrarch sought to look at them as emanations of a specific personality, they took on new importance. Petrarch tried to resolve this issue by assuming that the ancients' paganism sprang from ignorance. In the letters to Cicero and Livy he stressed that they had not had the opportunity to know Christ. In the case of ancients who lived after Christ the issue was more difficult to resolve, and Petrarch tried to gloss over it. Writing to Seneca, Nero's advisor whose spurious correspondence with St. Paul was still accepted as genuine in the fourteenth century, Petrarch noted that he could not tell whether Seneca knew of Christ or not (4:24.5). In Quintilian's case such ambiguity was less credible, so Petrarch chose to stress the personal connection between Christ and the reigning emperor, Domitian, signing his letter "in the 1,350th year of Him whom your lord preferred to persecute rather than to learn about" (4:24.7). Thus Quintilian occupied Petrarchan time indirectly through synchronism with an emperor whom Petrarch could see as a person but whom he did not need to respect.

Like its Augustinian counterpart, Petrarchan time had two dimensions. One was linear, universal, public, and all-embracing; the other episodic, immediate, private, and inextricably bound up with particular events. Petrarch's vision of the fall of Rome implied a linear time frame that embraced human culture from beginning to end with discrete periods marked off from one another. His understanding of the ancients as people, by contrast, suggested another kind of time, one having no particular direction or overall significance. It focused on individual relations and opened a way to understand the wisdom of others through the very communication among people that constituted the fabric of society. But this same communication, by creating a direct link with the ancient world, defied the linear time of general cultural change. Despite this tension, these dimensions were inseparable parts of a single time in the actual workings of events. The decline of Rome and the Renaissance of culture were too closely tied to specific events, and they lacked the transcendent will that made it possible for medieval theorists to ignore the particulars when necessary. Since no sophisticated humanist could completely dispense with either of these times, they coexisted uneasily in Renaissance narratives in a strained and richly productive symbiosis.

Petrarchan time did not replace either Hellenistic time, which measured the movement of empire, or the time of the incarnation, which measured the events of the Christian powers and the spiritual progress of humanity. Petrarch had little interest in the former, and he accepted Augustinian age theory as an overall structure for the latter. The fact that later humanists modified age theory shows that they considered it still a viable explanation of the temporal process to which it applied. Salutati, one of the most important followers of Petrarch, accepted the traditional picture of six ages, but felt that each age was itself a self-contained unit beginning with a creative event and ending with a calamity.[5]

As effective and generally accepted as it was, the time Augustine measured with age theory was not the same time that Petrarch had seen connecting him with the ancient world. In Petrarchan time no philosophical construct mediated the events of different periods. To portray events in that time Petrarch had to see simultaneously the differences and similarities between the ancient world and his own. Because of its stress on the actual context of individuals, Petrarchan time needed another measure, one that would be separate from both events it sought to connect, one that would be truly universal in that it was tied to no specific story. As humanists began to explore more fully the meaning of this time, the difficulty of measuring it became more evident. No existing chronology sufficed. The incarnational dating system was by then the standard for dating contemporary documents and was most practical, but in the fourteenth century it was too closely tied to the growth of the Christian commonwealth to be a universal dating system.

As long as Petrarch confined his attention to individuals, without writing a history of the fall of Rome and the advent of the Dark Ages, he could avoid the problem. But any attempt to connect the political life of the ancient world with that of contemporary Italy would bring it sharply into focus. To see the political culture in the same context that Petrarch had seen the individuals of the ancient world, scholars had to construct a measure for the time between the two ages. This proved a long and difficult task, undertaken under the stimulation of pressing political considerations. Toward the end of the fourteenth century Florence, under one of the few republican constitutions in Italy, found itself at war with the duke of Milan. In its diplomatic efforts to win support for its cause, Florence sought to present herself as the defender of liberty, and as part of that campaign leading scholars in Florence reexamined an important myth about the city's origins.

Until those wars with Milan, the Florentines had believed that the city was established by Julius Caesar, the founder of the Empire.[6] Such an origin was a badge of honor during the period when the Empire was regarded as the most perfect of all governments and as the means by which God unified the world and prepared it for the birth of Christ. An earlier generation of Florentines had been proud of this honor and had looked upon Caesar as a great and virtuous ruler, founder of the perfect state. Dante wrote in the early years of the fourteenth century that Christ's birth during the Empire proved God's favor to that institution,[7] and in the *Inferno* he condemned Brutus and Cassius to the deepest circle of Hell for assassinating Julius Caesar and resisting the coming of the Empire.

As the city's republican institutions became more of a factor in its struggle with Milan, Salutati, chancellor of Florence during much of the period, undertook a major reinterpretation of Florence's origins. He approached his task from a historical point of view. The idea that Florence was founded under the Empire was primarily theological in its orientation, though the chronicles recorded it as a legendary story. To portray Florence as the heritor of Rome's republican tradition, Salutati had to conceive the issue in historical terms and broaden thereby the scope of Petrarch's time. He extended the new time to express the historical context not simply of individuals but of whole societies.

The task of seeing Florence in this new time was more complex than Petrarch's understanding of Cicero. Salutati had to see not only the specific virtues of the Roman Republic; he needed to see Florence itself existing in the same time frame with the Roman consuls. In other words, he had to see the specific historical roots of Florence in a time outside that time dated by the incarnation, the chronology which had always measured the Florentine past.

In addressing this problem Salutati found that he had no means of measuring the time that connected Florence with its Roman origins. He could only present his new insight into Florence's past by treating the past and present in fundamentally different ways. To locate Florence's founding he subjected several classical authors to a philological analysis, building his argument in two steps. First he cited Sallust's *War with Catiline* and Cicero's second Catilinarian oration to prove that Catiline sent C. Manlius to Tuscany to recruit an army among those who were angry at being driven from their lands by the men whom Sulla had settled there.[8] The passages from Sallust and Cicero showed that Sulla had founded a colony in Tuscany, but how was Salutati sure it was Florence? The second step in his argument involved philological analysis of other passages in classical literature, including Florus' list of towns involved in the war between Sulla and Marius, a list which seemed to include Florence (pp. 30–35).

Salutati thus used no dates to locate the origins of Florence. He identified authors of acknowledged authority, calling Sallust a "historian of most certain truthfulness," and derived the circumstances of Florence's origin from their words. But it was not enough simply to identify the colony founded by Sulla as Florence. His essential aim was to connect this colony with the Florence of his own day, and to do this he had to describe the time that separated Sulla's soldiers from the fourteenth-century armies fighting with Milan. Here he discovered that, like Petrarch, he could not express this gap as a single period of time. Without giving even an approximate date for the colony, he went on to connect the Roman period with his own by dating the foundation of the priorate, the executive body which was a central feature of the Florentine republican constitution of his own day. "The fact that our city was ruled by consuls, as formerly Rome was, until precisely 1,282 years after the incarnation of Divine Wisdom, shows also our origins from the Romans" (p. 36).

In this passage Salutati continued his philological analysis to explain the continuity between the colony founded by Sulla and the republican Florence of the thirteenth century. He pointed to the fact that Florence's early officials, the consuls, had the same titles as the chief executives of the Roman republic. After noting this purely philological connection, however, he shifted into the incarnational dating system to express the priorate as an institution within that part of Florentine history that lay in the time measured by the chronicles and the public records. Between the Florence of the priorate and that of Sulla's soldiers there was no common measure of time. No single dating system available to Salutati could connect the two events without robbing them of meaning.

Salutati did not lack information about the period between the Roman Empire and the fourteenth century. Florence possessed a rich chronicle tradition that culminated in the work of Giovanni Villani, a Florentine merchant from the early fourteenth century who wrote a detailed account of Florentine affairs and whose work was available to the humanists of the period. Villani was spotty on details of early medieval history, but he did include them, and where he was wanting, Salutati had available a sufficient variety of early medieval historians, including Paul the Deacon on the Lombards and Jordanus on the Goths.

The information was accessible, but no one had conceived it as part of a distinct temporal sequence. Chroniclers such as Villani, who were not so interested in the conceptual structure of the past, used the two traditional dating sequences without any thought of integration. They recorded the events of Roman history as part of the sequence from the founding of Rome and the events of medieval history as part of the incarnational sequence.[9] To portray the events from Rome to the fourteenth century as part of a single time with both personal and social dimensions, scholars would either have to invent a new chronology or radically revise an existing one.

FLAVIO BIONDO AND THE DECLINE OF ROME

Narrative history presented special problems. Salutati, writing in the form of an invective, could use rhetorical devices to obscure the absence of a linear connection between the classical world and his own. By using philology to relate the consulship in thirteenth-century Florence to the Roman institution of the same name, Salutati made the connection through argumentation rather than through chronology. Such a solution would not work for narrative history, and the humanist historians who began to write narrative histories of their cities in the early fifteenth century found themselves faced with the task of constructing a chronology that would join these two periods of their past.

In a sense Salutati's work had complicated the task and made it more pressing. For Petrarch the change between Cicero's time and his was a cultural one. Rome was ignorant of Christ but rich in culture. The Dark Ages had gained knowledge of Christ but had lost the culture of Rome. His own age had before it the possibility of combining the gains of both ages into a new era. Thus the cultural change was paradoxical, but the paradox existed within a coherent framework and could possibly be resolved in the course of time. Salutati with his inquiry into political change had introduced a new dimension of human experience into the temporal process. Here too the relation between the past and present was paradoxical. The political change was both a loss of power and a transfer. The destruction of Rome's military might in the barbarian invasions went beyond the translatio imperii that Christian historiography had adopted from Pompeius Trogus, for no empire of Rome's stature had taken its place. The loss was as dramatic and unmistakable to the humanists as the loss of Roman culture. Yet there was a translatio, for the political units of the fourteenth century had acquired power, and this power was slowly becoming legitimate. Salutati began the work of analyzing this process, but to finish it his successors would have to tell the story of the translatio and place the political decline of Rome in the same chronology with the rise of the Renaissance city-states.

Leonardo Bruni (1369–1444), the first great humanist historian of the Renaissance, whose *History of Florence* was celebrated throughout Italy as a model for subsequent historical writing, attempted such a story. In the

first book of his history Bruni told the history of Tuscany from its beginnings until Florence won independence from the Holy Roman Empire in
1250. In that book he was faced with the necessity of describing these
trends within the new temporal perspective of Petrarchan time. Since
his work was a political history, he felt free to ignore the cultural changes,
but he still had to treat the invasions and the translation of power from
the emperors of Rome to the Holy Roman emperors of the Middle Ages.
But he did not present the invasions and the transfer of power as part
of a single trend. Instead he adopted two different chronological systems,
one describing the invasions and one the collapse of Roman political
institutions.

He located the first major invasion—the sack of Rome by the Visigoths—with reference to the founding of the city, thus using the dating
system common among Christian historians since Orosius. Bruni dated
subsequent invasions, however, from this first one. The Vandals, he said,
entered the city 43 years after the Goths; the Lombards 204 years after
them.[10] When he came to Charlemagne, however, Bruni introduced a new
dating system. With Charlemagne there was a clear change in polity, since
the Empire was divided between East and West and the novelty of the political situation in the West could no longer be ignored. To address the
new situation, Bruni interrupted his narrative, saying that he needed to
repeat a few things "in order to understand the matter better," and re-
capitulated the history of the imperial office.

In going over a period previously narrated as part of the barbarian invasions, Bruni introduced a new point of reference—the abdication of
Augustulus in 476 and the accession of Odoacer, the Gothic king who did
not assume the imperial title. Bruni claimed that with the accession
of Odoacer the imperial office ceased for more than 300 years—13 of
Odoacer, 60 of the Goths, 204 of the Lombards, and 25 until Charlemagne. Bruni thus portrayed the complex changes of the early Middle
Ages as two distinct sequences, each demanding its own chronology, each
with a separate reference point. In his narrative the process of integration
into a single time remained incomplete, though each of his two systems
did connect the ancient world with the early Middle Ages.

Bruni perceived the modern period of Florentine history as a coherent
temporal process, but he did so by ignoring its connection with the complex, separate processes he presented in the first book. He began his account of independent Florence without any dates at all, saying simply that
the events occurred "after [Emperor] Frederick's death" (2.27) and gave no
reference to the year of the incarnation until a decade of his story had
passed. Even though he conceived of the time that connected Florence

with Rome as linear, Bruni did not create a single linear sequence out of that period. He left those events in separate times, whose relation to one another was not brought into explicit clarity.

These writers of the early Renaissance had difficulty seeing a single lin-ear time linking Rome with contemporary Italy partly because they lacked a clear concept of the Middle Ages. Bruni, like Petrarch, saw the distance that separated the ancient world from his own, but he did not see distinc-tive characteristics in the period between the Roman Empire and Renais-sance Florence. The first historian to construct a narrative that portrayed the Middle Ages as a distinct period was Flavio Biondo, who imposed the division among ancient, medieval, and modern times that underlies our own periodization of Western history.

Biondo implemented his threefold division in a narrative of Italian his-tory from the end of Roman times down to his own. The narrative is di-vided into three parts. (Biondo began the fourth but died before he could finish.) In the first Biondo approached his task in a manner similar to Orosius' but with an essential difference. Where Orosius had told his story as the rise of Rome, dating from the founding of the city, Biondo told the story of its fall, dating from the city's first destruction. The event he chose was not the abdication of Augustulus in 476, which is the event com-monly used today to date the end of the Empire; instead he chose the sack of the city by the Goths. After dating this event with reference to the founding of the city (1164) and the incarnation (412), he went on to or-ganize the subsequent narrative by the number of years after the decline.

Just as Pompeius had seen the transfer of empire as a process which gave continuity to individual events, so Biondo conceived of the decline as a process, and used his dating system to connect concrete acts with it. "The Empire had already been declining [declinaverat] for three years when Honorius took hope from the dissensions among the invaders."[11] He also shared Pompeius' awareness that his principal theme was not the only pro-cess of relevance to his narrative. As the narrative progressed, other pro-cesses entered into the story, each requiring its separate time line. The decline of the Eastern Empire, which he thought began with the death of Heraclius in Asia, became part of the chronology. "It is now 275 years of of decline in the west and 76 in the east" (1.9). There is also a political process of decline as well as the military one that dominates the work. He dated the abdication of Augustulus on the same temporal line that Bruni had used, 517 years after Augustus. Finally he added the spiritual process of Christian governments. When the states that would rule in his own day entered the narrative, they occupied the time of the incarnation. Venice, he observed, was founded on the seventh day of the Kalends of April in

421 (1.3). Religious events, such as the arrival of a piece of the Cross in Rome, are also dated by the incarnation.

Even though mixed with other chronologies, the system of dating from the decline of Rome introduced important new elements into the narrative. Despite its similarities to the Orosian system, Biondo's differed in containing an implicit change of direction. Direction for Orosius was in any case not a major consideration in the actual narrative. Dating from the founding of the city served partly to order a series of catastrophes and disasters that had no particular direction. Even to the extent that the year of the city traced a transfer of power, that process had begun with Ninus and seemed to culminate in the world empire of the Romans. Biondo, by dating from the decline, brought a sense of rise and fall into the narrative of political events. The sack of Rome thus became an event that looked forward and backwards and described a change in direction. Biondo began the narrative by carefully describing the location of the event both with regard to his own time (1,030 years since the capture) and by listing the major periods of Roman history from the foundation of the city down to the sack (1.1). Moreover, he explained his chronology as a deliberate attempt to continue the dating system from the founding of the city, taking into account the new time. "Hence the computation of events is taken up from the founding of the city; we in a similar fashion have begun from its decline. By this means we can explain more easily and certainly in what order of times the events occurred and describe remarkable events which posterity scarcely believes" (1.1). The reference to the order of times is conventional, but it refers to the new time, the course of events from the decline of Rome, not the times of Bede or of Orosius.

Biondo also differed from Orosius in that his dating system began with a central event that was directly and internally related to the theme of his narrative. In describing and locating the events that followed on the sack of Rome, Biondo was recording the actual decline of the Empire. Though he did comment directly from time to time on the falling state of the Empire,[12] Biondo gave the reader far fewer theoretical digressions than had Orosius. He did not need to, for the progress of the decline was clear in the actual subject matter and was expressed directly through the chronology.

The reversal of direction described and dated in the first part of his narrative is an important stage in the development of a universal dating system, but Biondo never intended it to be universal. He dated the events in the second part, which recorded the beginning of legitimate political rule under the Holy Roman Emperor Charlemagne, from the incarnation. In that part, Biondo used the dates as he found them in his sources and gave dates frequently so that the reader could see clearly the temporal order and location of events. The second section of Biondo's work lacks the sense of direction that is so clear in the first part. One follows the events—limited largely to Italian history—from one chronicle to another, but Biondo never addressed the issue of how the decline was reversed, why it changed, or what significance the reversal had. His treatment in the second part stands in stark contrast to that in the first, where he explained and introduced the decline of Rome. Here he imposed little pattern at all on the events.

In the third part, which covered the forty years before Biondo wrote and which chronicled the events of the modern period of recovery, all dating ceased. The reader can no longer easily identify the precise temporal location. In place of the directional change of the first part, or the linear progression of events in the second, we find the alternation of peace and war. Biondo was more concerned there with the correct Latin for military terms and for titles than with the overall direction of events. He was still concerned with the order of the narrative, but now the order was philological rather than chronological. He began the third part by boasting that in the previous two he had reduced to order material that had been buried. The concern in the third part was no longer with temporal order, but with theme and philological accuracy.

It is not hard to understand the problems in Biondo's system. Part of the difficulty was purely technical. Since none of the sources dated from the year of the decline, Biondo had to calculate the dates of all events. This was not so much a problem in the early period, since the lack of a common dating system necessitated calculation in most cases, no matter what system was used. By the Carolingian period, when annals began using the year of the incarnation more regularly, recalculation seemed increasingly otiose, and Biondo himself began to use the year of the incarnation in preference to the year of the decline.

The year of the decline also presented conceptual problems. The perception of linear decline was more consistent with medieval historical attitudes than with those of the humanists. Gregory of Tours felt that he was living at the end of time and that his history was the record of a decadent and declining world. Biondo himself devised the periodization that contradicted this perception and saw humanity emerging from the Dark Ages into a new age of Renaissance. The dating from the decline was not a suitable means of conveying this attitude. It failed just as the important issue of transfer of power and culture to modern times arose. Biondo seems to have felt that neither his year of the decline nor the conventional year of the incarnation sufficed to convey the thematic realities of the history of contemporary Italy. Since Biondo's interest in contemporary history was largely philological, the reader is hard put to see an underlying temporal order in his narrative.

Even though Biondo's dating system had no followers and was used in only a limited fashion by its originator, it represented an important innovation in Western chronology. For the first time a chronology carried with it the notion of a reversal of direction. Central though the concept of reversal was to Christian historiography, medieval historians had used different and unconnected systems to portray the pagan and Christian eras. Since these systems measured unrelated times, they did not affect one another directly and could easily overlap. The system of dating from the decline of Rome was intrinsically related to that measuring time from the founding of the city. Decline presumed a rise. Moreover, the decline was tied to specific events more directly than either the founding of the city was to Orosius' theme or the incarnation was to the rise of Christian commonwealths. In those cases the system was more a convenience than a direct conceptual expression of the historical realities. By combining the dates from the founding of the city with the new measure from its destruction and decline, Biondo had created a system which was potentially all-inclusive, theoretically sound, and inextricably tied to the events and themes of an actual historical narrative.

ANNALISTIC FORM AND THE MEASURE OF PERSONAL TIME

Biondo created a linear sequence connecting the world of antiquity with his own, addressing the public and universal aspect of Petrarchan time. But Petrarchan time also created a more intimate, personal contact between the ancient and modern worlds, one in which the values and attitudes of daily life could be instructed and informed by contact with the ancients. Such contact would bridge the temporal gap between the two periods, ignoring the medieval period and making the chronology devised by Biondo irrelevant.

This personal time was also linear, but it retained episodic qualities that gave it a greater immediacy than the time measuring the history of the world into ancient, medieval, and modern periods. It was episodic in that it fostered direct contact between the individuals in modern times and classical exemplars regardless of the temporal distance between them. It was linear in that it saw past experiences as the building blocks of present realities. It saw the present not as a static illustration of the transcendental pattern of events but as a dynamic unity arising out of complex antecedent conditions. These conditions needed to be recorded and put in order before the strengths and weaknesses of the present day could be truly understood.

Because of its immediacy this time involved closer attention to the narrative structure than the other, and it conflicted more clearly with the traditional patterns of Christian historians. The Orosian picture of the past as a record of divine punishment for sin was firmy embedded in all of the narrative sources. Renaissance historians sought other lessons and imposed different patterns. They had no doubts about divine providence or the importance of God's favor. Biondo worked at the papal court, and even the most secular of Renaissance historians acknowledged God's direct intervention into historical events.[13] They doubted neither the power nor the existence of God, but they did reject the traditional model and drew another picture of the action of divine will in the historical process, one that looked more closely at secondary causes. This picture was ultimately more secular and fostered a more direct connection among concrete events.

This new insight into the structure of the past developed from several roots in the century after Petrarch, and it brought with it a new narrative technique to convey the new order. This technique organized events into a single linear series of years, where each year took its meaning from the events occurring within it. At the same time each year's events could be explained only by looking to the events of earlier years, and the years had a more direct relation to one another than in the Hellenistic annals. In this way the unity of the whole was maintained without sacrificing the importance of specific events within the time periods being measured. The historian who pioneered this method was Leonardo Bruni, whose narrative of Florence after 1250 contrasts sharply with his account of the earlier period in its structure and conceptual grasp of the issues.

Dominating Bruni's history is a picture of the overall meaning of Florentine history whose influence in the Western world has been decisive.[14] Bruni felt that Florence's greatness was a function of its republican institutions, that the competition fostered by these institutions unleashed and nurtured the psychological vigor of its citizens, motivating them to defend the city militarily with their arms, to beautify it with their patronage of the arts, and to enrich it with their energies. He correlated each of Florence's gains and losses with some change in its internal structure which either mobilized these energies or restricted them. At the beginning of his story, when Florence first acquired its independence from the Holy Roman Empire, he noted that the newly free people proved remarkably strong in fighting their neighbors (2.27–28). When the Florentines decided to elect officials by lot, Bruni complained that the procedure reduced the strength of the city by diminishing the zeal to excel that open competition for office had produced (5. 121–22).

Bruni offered his readers a picture of events tied together by a psychological relationship between individuals and their society. He followed the Orosian tradition in presenting the essential elements of the past as invisible and abstract, but he saw them as secular, tied to specific political decisions rather than to any general transcendental pattern in history. This momentous contribution had roots as complex as its influence was vast. In part it sprang from the same political experience that had inspired Salutati. Bruni had been chancellor of Florence for a decade when he began the history, a position which involved him in the foreign policy of the republic during its recovery from the bitter and protracted wars with Milan. During those wars, which were brought to an end only by the sudden death of Duke Giangaleazzo in 1402, the Florentines became convinced that their own obstinate resistance had saved all Italy from subjection to the tyrant. Furthermore, they felt their resistance drew its strength from their commitment to free, republican institutions. [15]

Pride in Florence's free institutions was nothing new in the early fifteenth century. Generations of the Florentine patriciate had vaunted them. What was new was the association of humanist scholarship with this commitment. Earlier humanists, including Petrarch, had few political interests and were even dubious about the advisability of leading an active life of participation in civic affairs. Some of the Florentine upper classes in the late fourteenth century even accused the humanists of lacking patriotism because of their unwillingness to help the city in its time of need. Bruni brought his humanist background to bear in the service of the city, and he searched the classical world for an expression of values that would be immediately relevant to his contemporaries. He found in Tacitus' *Histories* an attack on imperial government that became the foundation stone of his ideological support of the republic. He also discovered in Polybius' history of Rome the notion that the various elements of the past ought to be synthesized into a single view.

In one sense Bruni's view of the past was an extension of Petrarchan humanism. Like Petrarch, Bruni saw that there were differences separating the present from the past and that these differences involved values and attitudes. Bruni went beyond Petrarch to identify the actual human traits which expressed these differences and to establish rules for connecting these human traits with specific events of the past. Bruni had developed this theoretical structure some years before he began writing the *History of Florence*. Over a decade earlier he had already expressed these ideas in a dialogue in which he analyzed both the Roman and Florentine past. But to incorporate his theoretical structure into a comprehensive narrative of his city's past he needed a new measure of time.

The linear series based on the years of the incarnation, as it had been used in the Middle Ages, did not serve his purposes well enough. Even though his chronicle sources used it almost exclusively, Bruni seemed oddly reluctant, until the last section of his work, to date events by that system. His narrative of Florentine affairs began in relative time "after the [Emperor] Frederick's death" (2.27). Only one date appeared in the second and third books, which cover a period of 38 years, and his use of dates down to the period of the war with Giangaleazzo was sporadic. With few exceptions (1365, 1370, and 1380 are omitted), he dated at five-year intervals, adopting a practice common to many earlier historians and similar to that of Eusebius, who marked the years from Abraham and the founding of Rome at ten year intervals in his chronicle. At times Bruni dated political events, but without any consistent pattern and without indicating the most important events. He gave, for example, the precise year for the departure of Emperor Henry VI from Pisa but not for the death of Frederick II; he identified the year advisors to the priors were first chosen but not the year in which the priorate itself was created.

Instead of a linear series based on the incarnation, Bruni adopted the Hellenistic practice of annalistic organization. Two-thirds of his years were marked off and organized that way, and most of the longer periods not divided into individual years were treated as if they were a single year.[16] His most direct model for this practice was Livy, though he was well acquainted with Polybius, whose work he had translated into Latin, and with Tacitus, whose republican sympathies appealed to him.

Despite its advantages in clarity and comprehensiveness, the annalistic form did not serve Bruni as well as it had served the Hellenistic historians. Livy could start his year with an event that was intrinsically important— the consular elections—but Florence had no important annual political event, since its major officeholders were replaced several times a year. The spring military campaign began near the start of the Florentine year (March 25), and in the early books Bruni followed Polybius, using that military event for about half of his years. Unlike Polybius, he showed no reluctance to interrupt a single campaign to begin a new year, so that the annalistic organization took precedence over the continuity of the military narrative.

Military campaigns did not focus adequately on his central theme, the interrelation of psychological and political factors, and in the last half of the book the number of military introductions declined. In the last three books, which cover the war with Giangaleazzo, he introduced only one year with a military event. In place of the military campaign, Bruni began years with judgments on the psychological condition of the city at the time. "In the next year all was quiet abroad except for fear of the exiles"; "At the beginning of the next year great suspicion abroad and care for the affairs at home oppressed the citizens" (9.226,234). These psychological conditions could not be precisely located in time, for they refer to events which had already happened as well as to the background of those earlier events, calling readers to reflect on the whole history of the Florentine state as Bruni had told it in the previous books.

In the final books of the history Bruni described that bitter war with Giangaleazzo, duke of Milan, which had so deeply influenced his own intellectual development. In these books he introduced a new chronology, one which combined the linear series based on the Christian era with the annalistic structure of the Hellenistic historians. He dated every year by reference to the incarnation, thus creating a series of years covering the whole war but implicitly transcending the war and looking back to the series of preceding years already narrated. Within the period itself he divided his narrative into two parts. The first included the years of peace, in which he concentrated on diplomatic maneuvering and filled his account with references to the psychological state of the citizen body. The second part comprised the years of war, when he concentrated on military strategy and tried to explain the technical problems of the war. The two different types of years are separated by different vocabularies and different analytical perspectives as well as different subject matter. The years of war concentrate on strategic matters. Diplomatic events that occur during these years are ignored or glossed over; Bruni concentrated on the military power and size of the cities. In the years of peace he concentrated on diplomatic maneuvering, deemphasizing those military clashes that did occur. He stressed the psychological realities of diplomacy and tied them into the previously told story of Florence's growth through spiritual vigor. Despite this dichotomy between years of war and years of peace, the narrative attained unity by the coherent set of questions about the historical process which Bruni posed in the course of the preceding books. The episodic nature of the individual years acquired linear dimensions through Bruni's synoptical vision of Florentine history as a whole.

Bruni combined annalistic form with a linear series of years to present a new historical reality, one that constituted a process in time susceptible to analysis and capable of direct connection to particular events. Unlike Otto of Freising, Bede, or Orosius, Bruni had no need of abstract digressions to make his point; he could and must express it through the structure itself of the narrative. To do so he produced a new chronology. It was linear, in that it referred to a single process of Florentine growth, having a clear and observable beginning. The series of years from the incarnation, which Bruni found in his sources and used occasionally to provide signposts, expressed this aspect of the chronology, giving his work a linear quality that the Hellenistic narratives lacked. But the chronology was also episodic in that it invoked overlapping psychic realities. These realities took on clear form through the annalistic narrative which allowed concentration on a specific time period and presented clearly the interrelationships of the events occurring within each period.

Bruni, however, paid a price for this achievement. His narrative expressed the general, abstract realities that for him determined the course of Florentine history. He consistently avoided concrete detail and vivid imagery, lest they distract readers from the serious purpose of his history and obscure the important temporal processes. His work is intellectually satisfying but without color or passion. The narrative is rigorously and inflexibly selective, recounting only those events of the Florentine past that are relevant to the psycho-political state of the city. Finally, the very element of personality that underlay the Renaissance sense of time is absent from his narrative. Individuals have little role to play in the story. The general attitudes of the citizen body depend on institutional rather than personal factors. Individuals play the same role as they did in medieval histories; they are embodiments of specific virtues or vices.

Bruni's successors broadened his focus and addressed some of these issues. They looked for help to other classical models, especially Sallust, whose monographs were written largely without dates and who sacrificed chronological clarity to thematic unity when necessary. Lengthy digressions on military strategy or on the corruption of Rome interrupt a battle carrying the reader's attention to events far in the future, only to have the Sallust take up the story again from the point he left it without any indication to the reader that he has returned to where the battle ended.[17]

Sallust exerted considerable influence on Florentine historians of the mid-fifteenth century. Poggio Bracciolini, who succeeded Bruni in the chancery and wrote a continuation of his *History of Florence*, modeled his work on Sallust. Like the Roman historian he concentrated on war and diplomacy, treating the domestic history of the city only as it affected its military fortunes. Also like Sallust, he avoided annalistic form, narrating each war as a whole without marking the passage of individual years. In the early parts of the work, he marked the passage of one year into another only a third as frequently as Bruni, and even in the section on the war with Giangaleazzo, where he used Bruni as his principal source, Poggio indicated the change of years only half the time.[18]

Poggio's narrative is not without dates, however, for he used the year of the incarnation to mark the beginning and ending of the wars which constituted the substance of his narrative. He assumed behind the events a linear progression of years significant in itself. Unlike Bruni he did not see the year as an empty unit to be filled for the reader. Where Bruni occasionally told his reader that nothing important happened in a particular year,[19] Poggio always found some significant event to attach to any date he felt should be indicated, even if it meant going outside his scope to describe some domestic event.[20] Poggio used both Bruni and Biondo as models and sources, and his technique took the strong points from each to create a narrative that expressed Petrarchan time more effectively than either. The introduction of dates at key points gives the narrative a more linear quality than Biondo's, with its complete absence of dates during the modern period. At the same time the dates allow the particular events to express the essential structure of the past more effectively than Bruni's, with its dependence on annalistic form.

THE SHAPE OF PETRARCHAN TIME

The humanist historians faced three antinomies as they worked toward a narrative form for Petrarchan time. First, in the realm of subject matter public and private history each demanded attention. Public history required a narrative with linear and conceptual dimensions, while to express private history the humanists needed a more intimate and episodic approach that would convey their direct relation with individuals in the ancient world. Second, on a more abstract level the historians had to combine similarities and differences in their picture of the past. These could no longer be separated into such conceptually distinct categories as translatio imperii or metabole, which earlier historians had used. The meaning of the past required that simultaneous perception of similarities and differences which Foucault has called the hallmark of the Cartesian revolution and which depends on basic analytical categories, including time and space.[21] Since these categories lie outside of time itself, they allow an observer to bring together for precise comparison events widely separated in space and time. Third, continuity and change become part of the substance of time. This had been the case with medieval theory, but Renaissance historians embedded continuity and change in the very events that

give life to the story of the past. The unity went beyond theoretical speculation to inform the concrete events of the narrative.

These three pairs of opposites—public and private, similarity and difference, change and continuity—were too closely enmeshed to be clearly
separated so that they could be reconciled each in its turn. The existence
of one pair militated against a solution that would comprehend the others.
An abstract and absolute time line, unconnected with any particular story,
would be most fit for analyzing the similarities and differences between
various periods of the past, but such a time line risked diminishing the
personal and immediate elements of the past, elements which were essential to its usefulness in the eyes of the humanists. The actual chronologies
adopted by the humanists also sacrificed some elements and stressed others.
Biondo's time line measured change but could not easily convey the conti-
nuities with the past and did not function well as a vehicle for organizing
contemporary history. The annalistic form captured the relations between
events within an immediate temporal field but did not foster precise analysis of similarities and differences over long periods of time.

Ultimately the need to present similarities and differences won out decisively over the other conflicting claims and led to the establishment of
an absolute time line. But other solutions proved more attractive to the
leading Renaissance historians. Machiavelli and Guicciardini, each in distinctive ways, chose an approach which comprehended the personal and
private elements of the past without sacrificing the analytical perspective
on public life that had developed during the fifteenth century. These two
Florentines, close friends and active participants in the political life of
Italy, found history an important path to human knowledge as they tried
to understand their world. They lived in a time of devastating change for
Renaissance Italy. Foreign invasions by France and Spain, the collapse of
most regimes in Italy, profound social transformation, all contributed to a
sense that contemporary Italians were experiencing a degree of change un-
precedented in modern times. As Guicciardini described the first invasion
of the French, which began this period of crisis, "Charles entered Asti on
the ninth day of September of the year 1494, bringing with him into Italy
the seeds of innumerable calamities, of most horrible events and changes
in almost the entire state of affairs. For his passage into Italy not only gave
rise to changes of dominions, subversion of kingdoms, desolation of countries, destruction of cities and the cruelest massacres, but also new fash-
ions, new customs, new and bloody ways of waging warfare, and diseases
which had been unknown up to that time." [22]

In the throes of this catastrophe Machiavelli and Guicciardini disagreed
on the usefulness of studying the past. Machiavelli saw it as a repository of

examples to be emulated or avoided by his contemporaries. "The majority of those who read it," he complained, "take pleasure only in the variety of the events which history relates, without ever thinking of imitating the noble actions, deeming that not only difficult, but impossible; as though heaven, the sun, the elements, and men had changed the order of their motions and power and were different from what they were in ancient times."[23] In keeping with this conviction Machiavelli analyzed the events both of the ancient world and of Italian history to establish maxims which would guide the princes and republics of his own day to maintain their existence in a threatening world.

Guicciardini was more skeptical. He felt that the examples from the past were not directly useful, since history never repeated itself exactly and the minor variations could cause completely different results from courses of action that were ostensibly similar. "To judge by example is very misleading. Unless they are similar in every respect, examples are useless, since every tiny difference in the case may be a cause of great variations in the effects. And to discern these tiny differences takes a good and perspicacious eye."[24] Not only that, but change was so pervasive that all aspects of human conduct differed from one period to another. In an argument with Machiavelli over the similarity of the past and present, he wrote, "Given the variation in art, religion, and the changes in human affairs, we should not be surprised that there should also be variation in peoples' customs, which often take movement from institutions, chance, or necessity. It is thus true to conclude that the ancient times are not always to be preferred to the present, but it is also true that one age is sometimes more corrupt or more virtuous than another."[25]

Despite their disagreement, they developed a new sense of time, one that gathered together the elements of the different times used by their predecessors into a single framework that would be all embracing and would serve to explain and record the disaster that was taking shape around them. This time had a unity and comprehensiveness that no time before had manifested; it included the abstract and linear time of rise and fall, but it also expressed the episodic and personal time in which individuals could communicate with one another over different time periods. These historians conceived a time that stood unified behind any particular series of events and connected all possible periods, but they also saw a time that was inextricably joined to the particulars of their story. To achieve their vision and to reconcile the tensions in Petrarchan time, Machiavelli and Guicciardini used the notion of personality, connecting the individuals in their narratives more intimately with the analytical structure of the whole than had been the case in fifteenth-century histories.

In the hands of these historians the personal qualities of Petrarchan time were effectively extended to the political and social dimensions of the past. They produced narratives in which personal decision-making was a convincing part of the political process but where the vagaries of personal differences did not nullify the possibilities of analysis. This accomplishment had ramifications on both the theoretical level and the practical one. From a conceptual point of view, personal realities permeated the political theory of Machiavelli and served to reconcile the paradoxes that were implicit in Petrarchan time. In addition both writers implemented their theories in practice by developing narrative styles in which persons played a more effective role in the historical process than had been the case in earlier histories.

Some earlier writers had thought there were theoretical structures in the ancient histories, hidden to modern eyes, without which history would never achieve true usefulness,[26] but Machiavelli saw the problem in different terms. It was not lack of theory that vitiated the study of the past but failure to see it as a living being. "This neglect [of the past] is . . . due . . . to the lack of real knowledge of history, the true sense [senso] of which is not known, or the spirit [sapore] of which they do not comprehend" (*Discourses*, 1, Introduction). Machiavelli used terms of physical sensation here—sense and savor—to convey the quality of life and breath that the past must have if its students would truly understand it. History is not simply a repository of examples of abstract virtues and vices but a real thing, a series of real events that can be reproduced and given new life.

Machiavelli chose many ways to give life to the past, some of them literary, some conceptual. In one of his greatest works, the *Discourses on the First Ten Books of Livy*, he commented at some length on Livy's history of early Rome, hoping to use this important part of his heritage to understand problems of growth and decay in a state. He prefaced the commentary proper with a section of eighteen chapters in which he talked more generally about states without following the specific order of topics found in Livy. In these eighteen chapters, which he may have written earlier as a separate book, Machiavelli introduced his readers to a variety of perspectives on time, perspectives which give meaning and shape to the subsequent discussion but which cannot be reduced to a single philosophical point of view. Three basic approaches dominated his picture of political life: a linear one based on the physical existence of the city; a conceptual one based on the cyclic theory of rise and fall; and a moral one based on the pervasiveness of human evil. Machiavelli introduced his readers to these perspectives in the first three chapters of this early section.

In the first chapter of the *Discourses* Machiavelli explored the possibility that Rome's greatness could be traced to the circumstances of its foundation. "Those who read what the beginning of Rome was, and what her lawgivers and her organization, will not be astonished that so much virtue should have maintained itself during so many centuries; and that so great an empire should have sprung from it afterwards" (1.1). The key factor here is the fertility of the land on which the city is founded. Sterile land will create strong character but will not support a large population. Better to found the city on fertile land and use the laws to create the right attitudes. Machiavelli lay down here the parameters that determined the growth of cities in linear time, counting the time of the city from its founding, as Orosius had measured the translatio imperii. Moreover, he tied this time directly to the physical body of the state. The laws can give the state a distinct character but only within the limits imposed by its body. The state that Machiavelli will follow through the subsequent books of the *Discourses* is a physical entity with a definite and calculable beginning, existing in the sort of linear time that physical objects must exist in.

In the second chapter of the *Discourses* Machiavelli presented a different sort of time. Turning from the physical basis of the state, he considered its conceptual structure, its constitution. Taking as his guide the cycle in Polybius' sixth book, he explained how states pass in a cycle from each of the three good forms of government, monarchy, aristocracy, and democracy, into the three vicious ones, tyranny, oligarchy, and anarchy. From the constitutional point of view the state does not exist in linear time, with a definite origin exerting a continuous influence on its fate, but in a cyclical time where the origin is irrelevant and where a particular stage has no permanent importance, since it will soon pass into another. This aspect of the state is not timeless; it exists in time just as a virtue or a character trait does, but it does not exist in linear time. The origin of the constitution is not important, but the duration has significance in determining the outcome. "Such is the circle which all republics are destined to run through. Seldom, however, do they come back to the original form of government, which results from the fact that their duration is not sufficiently long to be able to undergo these repeated changes and preserve their existence" (1.2). Some states even escape the cycle for long periods of time. Rome, by combining the virtues of the three good types of constitution managed to survive for centuries.

There is yet a third time for states, a time which is neither linear nor cyclical but which lies in the wills of the inhabitants.

> All those who have written upon civil institutions demonstrate (and history is full of examples to support them) that whoever

desires to found a state and give it laws, must start with assuming that all men are bad and ever ready to display their vicious nature [malignità], whenever they may find occasion for it. If their evil disposition remains concealed for a time, it must be attributed to some unknown reason; and we must assume that it lacked occasion to show itself; but time, which is said to be the father of all truth, does not fail to bring it to light. (1.3)

Human will is intrinsically evil and needs constraint to prevent it from taking advantage of weaker elements in the state. By this secularized conception of original sin Machiavelli incorporated into the state an essentially Augustinian notion of time, one in which important realities are being revealed and where the structure of the state is a reflection of hidden desires and intentions. As an example of this time, he described the foundation of the tribunes as a necessary response to the arrogance of the Roman patricians, arrogance that had been kept in check by fear of the kings but which was unleashed with the expulsion of the Tarquins.

Just as the individual for Augustine possessed a body, intellect, and will, so for Machiavelli the state possessed physical, conceptual, and voluntary dimensions. It had a physical beginning and even an end. Machiavelli was ambivalent about the natural end of all states, but Guicciardini was quite convinced that states shared with individuals the fate of inevitable death. "All cities, all states, all reigns are mortal. Everything, either by nature or by accident, ends at some time. And so a citizen who is living in the final stage of his country's existence should not feel as sorry for his country as he should for himself. What happened to his country was inevitable; but to be born at a time when such a disaster had to happen was his misfortune" (*Ricordi*, 189).

The temporal aspects of the state do not overlap one another in Machiavelli's work the way different sequences did in earlier histories. Instead they become part of a single sequence which derives its unity from that of its subject. Any political event has for Machiavelli dimensions which are purely analytical, deriving from its constitutional structure, and which thus lie outside of any linear sequence. But these analytical elements have no intrinsic meaning apart from the particular physical and psychological dimensions of the state within which the constitution functions. Since political history represents a complex unity of analysis and particular physical events, it exists in a unified time. The linear order for Machiavelli possessed an intrinsic significance, meaningful in itself without external references, and differing significantly from that found in the looser narratives of previous historians.

The linear order of events in Herodotus, or even in Polybius and Thucydides, had no intrinsic significance, since it was often contradictory and contained more than one linear sequence, or even sequences in which the linear order did not exist. For Machiavelli this was no longer true; the linear sequence was intrinsically meaningful, and he used the chronological order of events as a locus and source of meaning in itself. When he explained the reason for Rome's early success, he discounted the notion that it was due to chance. He did this by analyzing the temporal order of Rome's early wars, showing that Rome never began one war before finishing the previous one. To make this point, Machiavelli had to assume an intimate connection between the intentions of the early Romans and the linear order of their physical actions. Since the wars occurred one after another, the Romans must have had an appropriate policy (2.1). This conclusion depended on the assumption of a single temporal field linking the intentions with the physical acts, just as Augustine assumed a link between intention and act in his own personality.

Machiavelli's sense of personal time determined the shape of his historical narrative. Because of the intimate connection between personality and politics, he introduced into his history of Florence a personal element that was lacking in both medieval and humanist history. Medieval historians tended to portray the individuals in their stories as examples of abstract character states, having little direct impact on the general process of events. Bruni paid little attention to the individuals and told his story as the interaction of abstract psychological states susceptible to clear analysis. Machiavelli made his individuals into living persons whose particular desires and actions had a direct effect on the process. He used a considerable variety of techniques to achieve this effect. He concentrated on specific details of people's behavior rather than on their general moral status; he described individuals in terms of their reputation, thus making their character an intrinsic part of the historical process, since it grew out of the perceptions of others. In addition he often gave differing and contradictory explanations for the conduct of specific individuals, making it hard to categorize people abstractly. Many of these techniques come from classical historians, especially Tacitus, but Machiavelli used them in a new way, to convey the personal reality of his characters and to give life to his narrative.

Machiavelli applied these techniques to bring alive the major characters of Florentine history. In the case of Michele di Lando he combined Bruni's interpretation with a variety of sensory details and concrete acts to produce his effect. Di Lando was a key figure in the Ciompi revolution, which threatened to destroy the basis of Florentine upper-class life in the late fourteenth century. Di Lando was the leader of the revolutionaries, but he deserted them at a key point and allowed control to pass back to the Florentine patriciate. Despite this service, most fourteenth-century chroniclers saw him only as the leader of a revolution they feared, and they portrayed him in the most vicious and pejorative terms. Bruni saw his significance and observed that the republic was saved only by Michele's virtue (9.224).

Bruni's reinterpretation of Di Lando was part of his general interest in political virtue. He had no interest in how Michele's personality had actually effected his result, and it is here that Machiavelli changed the material found in his model. In place of Bruni's abstract narrative of political events, Machiavelli gave Michele a physical presence and a set of clearly defined character traits. We see him entering the Palace of the Signoria in bare feet and later, as he began to accommodate himself to the established powers, drawing his sword in anger at a delegation of revolutionaries. Beside these concrete acts Machiavelli also provided a set of abstract characteristics, introducing Michele as one whose spirit, prudence, and goodness made him outstanding among Florentines. These abstract character traits do not stand alone, however, for Machiavelli gave them meaning through the concrete events of the narrative. Michele's acts were not conventionally kind or prudent, since he acted in uncontrolled anger and at one point sent a mob off to massacre an official. They were kind and prudent only within a specific context, at a particular time and place.[27]

Guicciardini, less convinced than Machiavelli that the past had useful lessons, made his individuals even more active participants in the historical process. Where Machiavelli stressed the action of individuals in political events, Guicciardini portrayed a more reciprocal relationship. In his *History of Italy*, which told the story of Italy's fall to foreign invasion, the reader can see individuals themselves change in response to particular events, just as their personal actions determine the course of wars and revolutions.[28] In Pope Julius II fear changed to anger as the actions of the French created new feelings. One provocation built upon another as his new feelings led him to actions that would not have been caused by his earlier state of mind. Julius' actions in turn created an entirely new diplomatic situation, including an alliance with Venice which seemed unthinkable in his first state of mind. The narrative is held together by the web of personalities that contended with each other for control of Italy. Analysis is intrinsically limited by the unpredictability and changeableness of these personalities. Yet analysis remains possible, since coherent personalities lie behind the changes. The whole of Italian history changes as an enlarged personality does—unpredictably but within general limits of possibility and character.

Machiavelli and Guicciardini conceived the past as a unified whole with many of the properties of the Augustinian personality.[29] Their vision brought together the antinomies of public and private, similarity and difference, continuity and change, that had stood apart in earlier Renaissance historians. This kind of time resisted measurement much as personal time had resisted it for Augustine, but the subject matter required a dating system, and they used the incarnational dating which by now was the standard in most sources for the modern period. In these narratives the year of the incarnation lost its last vestiges of ecclesiastical significance. Even Bruni had mentioned the Jubilee years—papal celebrations at the start of each century—but Machiavelli passed over the year 1400 without even noting the Jubilee until he had already told of the other events that occurred (3.29).[30] He mentioned the ecclesiastical ramifications of the year only if the events demanded it. "The year 1476 was passing and it was near Christmas, and because the prince [Galeazzo Sforza of Milan] usually went on St. Stephen's day in great pomp to worship at the church of that martyr, they [the conspirators] decided that this was the time and place to accomplish their design" (Florentine Histories, 7.24). Even Guicciardini, himself a diplomat in the papal service, mentioned the Jubilee of 1500 only in passing before telling of Caesar Borgia's capture of Forlì (History of Italy, 4.13).

Machiavelli, who devised the most comprehensive picture of political change since Pompeius Trogus, resembled the Hellenistic historian in his indifference to dates. He adopted neither annalistic form from Bruni and Tacitus nor the practice of dating significant events from Poggio and Sallust. The first dated event in the Florentine Histories is the Saracen sack of Genoa in 931, followed by the death of Henry I in 1024. Often where he found a precise synchronization of events in his source, he described the relationship generally as "about the same time."[31] Dates for Machiavelli had no independent existence. They functioned only to indicate and measure a continued process, and he tended to use them as much to express duration as to indicate a specific event. "Thus during the whole summer of 1483 things went well for the league" (Florentine Histories, 8.25). Even these dates tend to be haphazard; he felt no need to orient the reader in a numerical dating system. The general significance of the events lay in their internal temporal relationship, not in their synchronization with an external time line. The dates thus became a part of the narrative, rather than standing outside it; the narrative drew its order and meaning from the events. "The war in Tuscany had already lasted almost a year and the time had come in 1453 for the armies to take the field, when Alessandro Sforza, the duke's brother, came to the aid of the Florentines (6.30).

Guicciardini, more concerned with the particular than Machiavelli, had greater need of an explicit chronology. He chose to organize his work annalistically and, like Bruni at the end of his history, tried always to combine specific events that could be precisely located in time with assessments of the general significance of the process created by those events. As part of his concern to keep the events as the central focus of his story, he made sure that each year was marked by some datable occurrence, even though to explain its significance he often had to discuss psychological and strategic considerations which went beyond the year in question and gave the reader a sense of the general ramifications. Often he began a year with a discussion of its general significance, forcing the readers attention forward and backwards in time, only telling of the datable occurrence that marked the year at the very end. "These things were done in 1500. But much more important things were ordained for 1501 by the king of France, to be more free for which he reached an agreement with the king of the Romans." What actually occurred in that year was a truce with the emperor, but Guicciardini followed this introduction with a full discussion of King Louis' plans for recovering Milan by marriage, plans which dated back several years. Only then did he explain that "in the beginning of 1501, he obtained a truce for several months from Maximilian" (*History of Italy*, 5.3).

The years seem to take on a life of their own, especially if they were unusually significant or disastrous. The year 1527 was one such year. Imperial armies out of control ravaged Italy, invaded Rome, and put the city to the worst sack it had ever endured. In his account of that year Guicciardini assessed the significance of the disaster, drawing the reader's attention to the general causes and blunders that had brought it about, but also tying it to the specific problems of undisciplined soldiers that were its immediate cause. "The year 1527 will be full of the most atrocious events, unheard of even for centuries. There will be changes of state, evil on the part of princes, horrible sacks of cities, great famines and diseases through almost all of Italy. All will be full of death, flight, and rapine. To which calamities no difficulty was more pressing to begin than the difficulty that the Duke of Bourbon had in moving his Spanish infantry from Milan" (18.1). The specific event that begins the year foreshadows the general problem; the duke of Bourbon's inability to discipline his troops was only the first sign of a process which saw drunken Lutheran soldiers shooting at Pope Clement as he fled the Vatican for the safety of the Castel Sant'Angelo. At the same time the introduction places the events of 1527 in context with the whole series of events that led to the sack of Rome. Moreover, it looks beyond the war itself, to the constant issues of princely vice, greed, and stupidity that brought Italy to its sad condition and that had plagued other states in similar ways. The year contains similarity and difference, constancy and change, private and public sorrows.

Guicciardini created in the *History of Italy* a linear series of years that provided a more explicit and forceful continuity than that of Tacitus or Thucydides, whose histories he greatly admired and whose narrative techniques he imitated. He endowed the years of the incarnation with a vitality and immediacy which the consular years of the classical historians lacked. To do so he gave them personal dimensions they lacked in either Christian or earlier humanist historiography. The year became the medium between personality and political change, giving to Guicciardini's narrative its dynamic quality.

He used the year to characterize political change at the very outset when he described the coming of the wars to Italy after a long period of peace and tranquillity. "It is clear that since the Roman Empire, weakened principally by the decay of her ancient customs, began more than a thousand years ago to decline from that greatness to which it had risen with marvelous virtue and fortune, Italy had never known such prosperity nor experienced such a desirable condition as that in which it safely reposed in the year of Christian salvation 1490 and the years just before or after" (1.1). The reference to the decline of Rome and the rise of the Italian city-states makes this passage part of the tradition of humanist historiography, but the concentration on the year 1490 is strange, for he is referring to a considerable period of time, not simply that year or even the years immediately before or after. Furthermore, he had in mind no specific event that occurred in 1490, for he went on in the passage to characterize the state of Italy in quite general terms without reference to anything that had happened in that year.

The year 1490 stood in Guicciardini's mind not for a particular event but for a person, Lorenzo de' Medici, whose virtues and diplomatic skill had played such a role in creating this state of peace. "Many causes preserved Italy in this happiness, which had been attained by various circumstances, but among these by common consent no small praise was given to the industry and virtue of Lorenzo de' Medici." This comment is of course partisan and reflects Guicciardini's ties to the Medici, but it has a deeper significance, for it connects the general state of Italy with the physical and moral existence of an individual. Guicciardini reinforced the point by going on in the passage to tie in the other personalities who dominated the affairs of Italy during that year, Pope Innocent VIII, Ferdinand of Aragon, and Ludovico Sforza. Guicciardini admitted freely that the rulers were each pursuing different interests, but despite these differences the tranquillity continued, because Lorenzo, Ferdinand, and Ludovico had a common desire to maintain it. "Since there was the same will for peace in Ferdinand, Ludovico, and Lorenzo—partly for the same and partly for different reasons—it was easy to maintain an alliance."

By connecting the state of peace in Italy with the individuals whose wills preserved it, Guicciardini set the stage for the dramatic change into war, for he tied political changes to personalities and made them one with the facts of birth and death that are the essential features of individuals.

> Such was the state of things, such the foundation of the peace of Italy, so disposed and balanced that not only did no one fear any present change but one could not easily imagine what plots, events, or arms could disturb such tranquillity. Then, in the month of April, 1492, there occurred the death of Lorenzo de' Medici. . . . The death of Lorenzo, as things prepared every day for the future calamity, was followed a few months later by that of the Pope. (1.2)

The death of the pope brought another personality to dominate and characterize the political and diplomatic life of Italy.

> For Alexander VI (as the new pope wished to be called) was of singular wit and sagacity, excellent counsel, a marvelous capacity to persuade, and with great resourcefulness and incredible energy to encounter any happenstance; but these virtues were far outweighed by vices: the most obscene habits, neither sincerity nor shame nor truth nor faith nor religion, insatiable avarice, immoderate ambition, more than barbarous cruelty and a burning desire to advance in any way he could his children, of which there were many. (1.2)

Guicciardini described the change from a happy state of peace to a bloody state of war and dissension as both a change of personality and a progression of years. The year 1490 represents the period characterized by the benign Lorenzo and 1492 is the year of the vicious Alexander. By conceiving the years as if they were proper names and associating them with personalities, Guicciardini conveyed the essence of the change through the linear progression of years. Thucydides used a different time frame to describe the period of peace in which Athens grew to power from that he used for the outbreak of war. Polybius used a different time frame for the period before the integration of the world by Roman conquest from that he used for the Oecumene. Guicciardini used a single time frame to describe the passage from peace to war, a time frame that took its reality from the personalities whose birth and death gave an intrinsic sense of change and linear order to the political events of the Italian wars.

Throughout the work, years take on personalities and characteristics of their own. They have not simply the abstract qualities that Hellenistic historians ascribed to them but dynamic and concrete features. "With people's spirits in this mood and with such confusion that things were obviously moving toward new troubles, the year 1494 began, . . . a most unhappy year for Italy, and truly the first year of the years of misery, because it opened the way for innumerable horrible calamities of which one can say that for various reasons they affected much of the world" (1.6). Here again Guicciardini marked the year by a change in personalities, for he passed on from this introduction to discuss the French king's rash dismissal of the Neapolitan ambassadors and the subsequent death of Ferdinand of Naples, whose sagacity had kept the unstable alliance of Florence, Milan, and Naples in order.

The personal reality of life and death which Guicciardini embedded in his narrative was more than an abstract construct. It was a fundamental fact of existence, beyond full explanation, and most effectively portrayed through the story. Guicciardini made the year itself into a person, with not only a clear beginning and ending but a name. By so doing he made personality a metaphor for time not only in theory, as Augustine had done, but in the structure and substance of the narrative. The story itself was composed of discrete, individualized temporal units that expressed judgments which Guicciardini made on the collapse of Italy during the sixteenth century.

Since the year itself took on the reality that Western thought had long accorded personality, events in different theaters existed in the same time and their correlation comes to be expressed with a new precision. Guicciardini was careful to tell his readers which convention he used to mark the beginning of the year, since the new year was the birth of a new being (1.6), and he used the exactitude this brought to reinforce in our minds the sharp and sudden beginnings and endings which punctuate his story. At one dramatic juncture the French king made his way to Rome, amidst frantic negotiations with the pope over the invader's role in forcing out the Neapolitans who were occupying the city. Guicciardini described the resolution of these negotiations in a way that brings the year and the theme of sudden endings and beginnings together with vivid impact.

But finally, deciding that this was the lesser of all his dangers, he [the pope] accepted these demands and made the Duke of Calab-

ria depart from Rome with his army, first having obtained from
Charles a safe-conduct so that he could pass safely through the
whole territory of Rome. But Ferrando, having magnanimously
refused it, left Rome by the Porta di San Sebastiano on the last
day of 1494 at the very hour when the King was entering with the
French army by the Porta di Santa Maria del Popolo, again armed
with lance on hip, as he had entered Florence. At the same time
the Pope, full of incredible fear, dread, and anxiety, had retired
into the Castel Sant'Angelo, accompanied only by the cardinals
Batista Orsini and Ulivieri Caraffa, a Neapolitan. (1.17)

Guicciardini built his story on simultaneous beginnings and endings correlated with the change of the year. Time existed in the narrative not as an
external series of numbers but as a series of intervals, with discrete points
that marked these intervals and made possible precise judgments about
temporal relations. The synchronisms that emerge form the events in different parts of Italy into a linear series of events, interacting with one another in complex ways, but all comprehended in a single temporal sequence. The time line is universal, comprehensive, and linear but has no
existence outside of a particular narrative. It depends on the concrete
changes to which it gives form and meaning.

Petrarch's longing for a personal relationship with the ancients produced a new sense of time. To a certain extent it implemented important
aspects of Augustinian time which had remained only theoretical in medieval historiography. Personal will became an integral part of the historical
process, and time became personal on a political as well as on an individual level. To incorporate personal will into a meaningful political process
Renaissance historians had to ignore the teleological quality of Augustinian time. History has no external direction in these narratives; human will
functions only in immediate situations with specific goals. The sense of a
single goal toward which all human history is moving—though accepted
in theory by the humanists—is not an important part of their historical writing.

No single approach could incorporate all the dimensions of Petrarchan
time. Machiavelli and Guicciardini succeeded remarkably well in combining the political and personal elements of the past into a single story.
Their stories also convey in a single gesture the dimensions of continuity
and change that were essential to the humanist perspective on the past.
But their very preoccupation with the particular and their use of the concrete event to convey the relationship between the individual and society
made it difficult for them to see the past in terms of precise differences and

similarities. These can emerge more clearly from an external perspective which constitutes an abstract structure without specific content. To classify and study similarities and differences, the historian would have to stand outside the societies being studied and apply absolute standards of comparison to them. Such an approach became the dominant theme of the late sixteenth and seventeenth centuries. Since it produced a remarkable revolution in science and thought, leading to the work of Descartes and Newton, we tend to consider it the more advanced and productive of the possible tendencies in Petrarchan time. But to implement an abstract and scientific history, historians had to surrender the attention to the personal realities of the past that gave to the works of Machiavelli and Guicciardini their life and their impact. The gains in precision that will be the subject of the next chapter had their price. They came at a cost in breadth and immediacy.

7

The Dating of Absolute Time

In the beginning God created Heaven and earth which happened at the beginning of time (according to our chronology) in the first part of the night which preceded the 23rd of October in the year of the Julian Period 710.

James Ussher, *Annales Veteris Testamenti*
a prima mundi origine deducti (London, 1650)

The Times are set down in years before Christ.

Isaac Newton, *The Chronology of Ancient Kingdoms Amended*
(London, 1728)

Bishop Ussher is famous among students of geology and evolution as that seventeenth-century divine so confident of his chronological technique and so convinced of the literal truth of the Old Testament that he calculated the exact time when God created the earth. The bishop was not alone in his presumption. Though few of his contemporaries were quite so precise in their judgments, they would have agreed with him in his general assessment of the age of the earth and in the possibility of dating it with accuracy. Ussher was extreme for his age, but the age itself was unusually sure of its dates.

The controversy over evolution has made Bishop Ussher into a symbol of narrow religiosity and opposition to scientific progress, but his confidence in the dates did not spring from religious enthusiasm. It reflected

instead some of the scientific trends of his day. In particular, two important developments of the sixteenth century made scholars more sure of
their dates. First, the century witnessed the creation of a new historical
methodology that employed the critical techniques of Renaissance scholarship to assess the reliability of ancient historical sources. Second, a new
dating system, that of the Julian period, based partly on this methodology,
made it possible to integrate events in all time frames into a single system.
Bishop Ussher in fact used this system in preference to the B.C./A.D. one,
including those dates only for the convenience of his general readership.

The new methodology and the Julian period were crucial to the forma-
tion of a dating system for absolute time, but they did not immediately
create such a system. That system, based on the years before and after
Christ, came into general use only toward the end of the seventeenth century with chronologers such as Isaac Newton. Its usage made possible a
new sense of historical narrative, one which conceived time frames as
empty units, which were devoid of particular meaning but which gave substance and form to all events.

Enlightenment historiography is replete with references to time that assume such a continuous and complete time line. Voltaire began his *Age of
Louis XIV* by claiming that he was going to write of "the spirit of men in
the most enlightened age the world has ever seen."[1] To substantiate this
claim, he surveyed briefly the previous ages that could pretend to some
degree of enlightenment: classical Greece, Augustan Rome, and the Renaissance. "Every age has produced its heroes and statesmen; every nation
has experienced revolutions; every history is the same to one who wishes
only to remember facts. But the thinking man and what is still rarer the
man of taste, numbers only four ages in the history of the world; four
happy ages when the arts were brought to perfection and which, marking
an era of the greatness of the human mind, are an example to posterity."
Voltaire saw enlightenment as a historical entity, existing and changing in
time but transcending the time frames that had previously measured particular societies. Time for Voltaire applied to and integrated all epochs; it
was a means of evaluation depending on a chronological schema which
put all events of human history into a single time frame. Such a schema
would allow universal comparison and produce Voltaire's comprehensive
judgment that only four enlightened eras had existed. Gibbon displayed a
similar time sense when he opened his *Decline and Fall* by assessing the
civilization of Rome in the second century of the Christian era and later
calling that period the happiest in the history of the world.[2] The time
sense of these enlightenment historians differed markedly from those of
earlier ones and should be understood against the background of the

changes in chronology and science in the seventeenth century that produced the modern dating system.

THE BEGINNING OF ABSOLUTE HISTORY

Machiavelli and Guicciardini approached time from a literary rather than
a technical point of view. They had little interest in chronology as a scientific discipline and saw temporal relations in a different context from those
who sought to define them abstractly through a numerical system. For
those historians of the Italian Renaissance who did try to present the relation with the ancient world in numerical terms there were two principal
sequences to historical time, roughly defined as ancient and modern. They
counted ancient time in years from the foundation of Rome; modern time,
in years after the birth of Christ. The time of Rome recorded the growth
and decay of empires; that of Christ, which told of the rise of the Christian
commonwealth for medieval historians, became the story of the growth of
independent city-states, as that time took on increasingly secular qualities.
Scholars might disagree about the precise relationship between ancient
and modern times, and the fall of Rome never achieved satisfactory integration into one of the dating systems. Nevertheless, the humanist historians of Italy, like their medieval predecessors, were satisfied that they
could effectively place all important events in human history in one or
both of those two sequences.

Several important developments in the mid-sixteenth century threatened this confidence. Cultures appeared that were previously unknown to
Europeans, while others never before conceived as part of a single historical process became too important to ignore. On a deeper level, a new
sense of universal history arose, one that saw all events of human history
as part of a single process. On a theoretical level this sense was less universal than Augustine's, since it was circumscribed by the invention of writing, but on a practical level it was more comprehensive, since it suggested
the importance of placing all historical events on a single time line that
would embrace the whole of human history in one linear series. In addition to these changes in historical consciousness, the heightened interest in source criticism that characterized the Renaissance created new
standards of truthfulness, which in turn fostered a need for more precise dates.

The appearance of new cultures was particularly threatening to the established chronology. The voyages of exploration brought Europeans into
closer contact with the old cultures of Asia and exposed them to the com-

pletely unfamiliar states of the New World. The Inca and Aztec empires could not be effectively integrated into the history of the West as long as the time sequences were tied to themes with specifically European content. To construct a universal history that truly comprehended all known empires and peoples, scholars would need a broader measure of the whole reach of historic time. In addition to the discovery of new empires, events within the West dispelled the illusion that the two sequences were adequately comprehensive. The rise of the national cultures of France, Germany, England, and Spain led scholars to recreate the medieval past in more positive terms than those used by the early Renaissance writers. Franks, Goths, even Huns, made positive contributions to European society. Their story demanded an attention that would not submerge it under the tale of Rome's rise and fall.

Even among those events that had always fit in the ancient time frame, historians began to see problems, especially with the integration of sacred and secular time. Here historians began to wonder seriously about the truthfulness of their sources and the adequacy of the traditional means of validation. Could they really claim a universal perspective on the past when so many of the events recorded by ancient historians were unknown to the Hebrews, and the ancient historians seemed ignorant of such important sacred events as the Exodus or the Flood? Since no ancient source was conscious of all events in the past, the sixteenth-century historians felt they must reconstruct a universal picture for themselves. Formidable difficulties lay in the path of such an endeavor, particularly in the matter of validation. Did the absence of an event in one historian's writings mean that he was ignorant of things going on outside his immediate frame of reference or did it mean that the event never occurred? Which sources were reliable and which were not?

The issues of comprehensiveness and validation were closely connected. As writers in the late sixteenth century felt more acutely the problems of integrating a complex variety of events drawn from independent sources and as they worked more actively to create a history that was truly universal in scope, they saw clearly the need for a new method of validation. They had to establish a system for deciding in principle which sources were accurate and which were not. Only thus could they create a history that would include all possible events, those now part of the historical record and others that might come to light in the future in sources not yet discovered.

To test the validity of their sources, scholars borrowed techniques from the law, where the Renaissance insight into the historical context of ideas was causing important changes. Medieval jurists had looked on Roman law as a coherent body of legal principles. By the sixteenth century humanists realized that the basic Roman law code, the Pandects of Justinian, was a collection of mutually contradictory laws coming from different periods of Roman history. Roman law was positive law, representing the acts of specific political bodies, not a coherent body of natural law, deduced from abstract and self-evident principles of government. In exploring the significance of this insight, humanist scholars arrived at a methodology which brought time and truth together. Guillaume Budé, one of the leading early French humanists, analyzed the Pandects in 1508, showing that the conventional texts were corrupt. They had, moreover, often been misinterpreted by legal scholars who ignored the particular qualities of classical culture which had given meaning to the original laws. Most important, Budé showed that since the Pandects embodied the history of Roman law, their chronological order had an intrinsic significance. Only by ordering the edicts in time and seeing the development of Roman law against the background of changing problems that it was designed to solve could one see their true meaning and find the legal principles that had guided the Romans.[3]

Budé had shown that Roman laws were historical and contingent, but the ideal of a coherent, universal series of legal principles did not perish. Instead legal thinkers wondered if the myriad laws of all the legal systems that humans had lived under in the past might embody such principles. Among the many thinkers who addressed this problem of finding a universal law, the French were particularly active. During the 1560s Jean Bodin, among several of his countrymen, came to the conclusion that Roman law could not be used as the basis of a legal system. They felt that we could understand the basis of statecraft only by seeing all states in their historical context.

Any new perspective would involve a close, comprehensive, and systematic study of the past, aimed at creating a single set of laws that transcended any particular legal system. To achieve the desired goals, Jean Bodin devised a new method of history in his *Method for the Easy Comprehension of History*, published in 1566. Essays on the art of history were common in the sixteenth century, but Bodin's work differed substantially from earlier ones. Traditionally they had focused on such literary issues as style and tone or on such moral issues as the need for honesty. Some had identified the topics worth including in a history. Bodin concentrated on the issue of truth and asked how to validate the events of the past as they were found in traditional historical writings.[4]

Bodin found that the rise of skepticism during the previous generation had exacerbated the problem of validation. Many in the sixteenth century reacted to the confusion and uncertainty of conflicting sources by distrusting all authority. These scholars maintained a total skepticism about the past. Bodin realized that such an attitude robbed history of all usefulness just as surely as excessive credulity robbed it of its truth. "If we agree to everything in every respect, often we shall take true things for false and blunder seriously in administering the state. But if we have no faith at all in history, we can win no assistance from it." [5] Bodin began by offering his readers some guidelines for steering a course between credulity and excessive skepticism. We should value historians who have attained their information firsthand, distrust those from societies which suppressed free expression, look for writers who have had some administrative experience, and have confidence in those that have been universally praised (pp. 47,46,50,56). To help his readers apply these principles Bodin included in the *Method* appraisals of specific ancient historians.

These guidelines alone could not assure a true understanding of history. To use history effectively, to construct from it universal principles of statecraft, a deeper and more comprehensive perspective was necessary. Proximate judgments about the honesty and reliability of particular historians were not enough. We must tie our judgments about the particular events to a sense of the overall pattern of history and find this general pattern in the unchanging regularities of the natural world.

> Since, however, the disagreement among historians is such that some not only disagree with others but even contradict themselves, either from zeal or anger or error, we must make some generalizations as to the nature of all peoples or at least of the better known, so that we can test the truth of histories by just standards and make correct decisions about individual interests. . . . Let us seek characteristics drawn not from the institutions of men, but from nature, which are stable and are never changed unless by great force or long training, and even if they have been altered, nevertheless eventually they return to their pristine character. (p. 85)

Bodin thus went outside historical time to find regularities in nature itself. The laws he found were geographical and climatic. He constructed a system by dividing the earth into three major climatic zones—northern, southern, and temperate—each fostering distinctive psychological, physical, and moral features in their inhabitants. The system is essentially a spatial one, though there are occasional attempts to introduce a temporal dynamic. He thought the history of the world could be divided into three 2,000 year periods and felt that important changes in world history could be correlated with Pythagorean numbers (pp. 122,223–36). His over-whelming concern, however, was to correlate the specific events and characteristics of past cultures with their spatial location.

This conception of time resembles the Augustinian perspective in that it looks at all historical events from a vantage point outside of the events themselves. There are of course important differences. Bodin was more secular and looked at the past for political rather than religious insight, valuing the pagan historians as much as the Scriptures.

> Since the system of governing a state, knowledge, and lastly civilization [humanitas] itself have come from the Chaldeans, Assyrians, Phoenicians, and Egyptians, at first we shall study the antiquity of these races, not only in writers who have written of them especially . . . but also from the Hebrew authors, whose affairs have much in common with the rest. . . . Afterwards we shall investigate the history of the Hebrews, but in such a way that we shall study at first the system of establishing a state rather than a religion. (pp. 22–23)

The epistemological differences between Bodin's universal history and Augustine's are equally important. Augustinian time had theological sources, which made it a pure theory that proved difficult to relate effectively to the concrete events of the past. Bodin, by contrast, drew his model directly from empirical data, deriving much of its verisimilitude from a detailed analysis of particular cultures. The general summary of his position here gives a quite mistaken impression about the degree of generalization that characterizes the book. The *Method* is replete with minutely detailed discussions of particular societies and laws; only occasionally did Bodin rise above this detail to make his generalizations.

The empirical derivation of his model opened the possibility of a new approach to historical knowledge, one which would integrate the methods of source criticism into theories of historical change to claim accuracy, usefulness, and comprehensiveness for its perspective on the past. Bodin's *Method* began the task of achieving an integrated methodology, but some important epistemological problems remained, especially in a period not yet in possession of the concept of an absolute subject as the basis of objectivity. How could the method prove its own validity and substantiate its claim to present the objective past?

If the focus of study could be shifted from history to historians, and if the changes in human society could be correlated with changes in historical method, then these problems could be solved, for then the method could use itself as the basic subject and establish its validity internally from its own sources. In one of the earliest attempts at a systematic historiography, La Popelinière wrote around the turn of the seventeenth century a *History of Histories* in which he not only included a compendium of historians but suggested a periodization for the development of historical writing. Historical writing, he was convinced, had a similar beginning in all human societies.[6] From this beginning, which he called natural history, history changed into poetry, as each society learned to write; then came a stage of prose annals, and finally the fourth stage of mature history, associated in the West with the writing of Herodotus. At all stages historians depended for their values and questions on the society they lived in. Critical examination of these values gave insight into past societies through the very writings by which they were known to us.

The insight that the writing of history itself exists in time led La Popelinière to insist even more strongly than Bodin that history must be universal. In a later work, he criticized all previous historians for writing particular rather than general history, for not leaving the immediate context of their subject. The very insistence on writing of events they had witnessed or for which they could interview witnesses led to these limitations and condemned older histories to narrowness. "Under those conditions you would always be ignorant of the past . . . you would only have particular histories which would give you information only about events which occurred in the writer's presence."[7] Instead of such particular studies La Popelinière urged the writing of general history. By that term he did not simply mean that history should be universal in subject matter, but that it should be general in method, that it should consider all aspects of the society. The perfect history includes the "total and complete substance of the states it wants to represent."[8]

In the work of Bodin, La Popelinière, and other French scholars the field of historical inquiry was broadened to include all possible events in any culture and all aspects of a society, including its historians. The time within which the absolute history occurred was essentially and unconditionally outside of the field of any historical event, including the very recording of historical events that made the study of history possible. The meaning of the historical events depended on their place in the time line, and the time line antedated both the events and their historians. Such a concept of history was absolute—La Popelinière called it "accomplie" or complete—in the sense that it was conceptually prior to any particular series of historical events and comprehended all possible events that had been written down. No existing chronology could serve to implement such a time line. Absolute history would require another dating system, one that began before any historical event and continued far enough in the future to include all events as part of a single time.

THE JULIAN PERIOD AND THE NEW CHRONOLOGY

Bodin was acutely aware that his method gave a new urgency to the problem of dating.

> Those who think they can understand histories without chronology are as much in error as those who wish to escape the windings of a labyrinth without a guide. The latter wander hither and thither and cannot find any end to the maze, while the former are carried among the many intricacies of the narrative with equal uncertainty and do not understand where to commence or where to turn back. But the principle of time, the guide for all histories, like another Ariadne tracing the hidden steps with a thread, not only prevents us from wandering, but also often makes it possible for us to lead back erring historians to the right path. . . . Since the most important part of the subject depends upon the chronological principle, we have thought that a system of universal time is needed for this method of which we treat. (p. 303)

Bodin's chapter on chronology in the *Method* reveals clearly the obstacles standing in the way of scholars during this period. First, they would have to establish a suitable starting point. The founding of Rome, or of any other empire, would not serve to orient a truly universal chronology. Something more comprehensive and outside of any particular event would be far better. The creation was conceptually suitable, but both metaphysical and practical difficulties stood in the way of its use. Bodin devoted the first part of his chapter to a discussion of whether the world had a beginning. Since that issue could not be resolved empirically, he had to approach it as a problem in metaphysics and deduction. "Since, then, the celestial matter is flowing and transitory, we infer that it had a beginning and will sometime end. This being the case, it is evident that time has a beginning and an end" (p. 319). By this discussion Bodin gave a sound foundation to universal chronology. But no practical system of chronology could rely on pure deduction; it had to be grounded in historical events. "Now our system of chronology from the Creation must be taken out of historical documents" (p. 319). The rest of Bodin's chapter is devoted to minute analyses of discrepancies in the sources. Though he did offer some opinions as to which was right in particular circumstances, Bodin did not create a general approach to chronological technique. His goal was primarily to acquaint his readers with the major opinions, not to give them a method for reconciling them. The chapter concludes not with a system of chronology but with a disquisition on the end of time.

La Popelinière, starting his system of absolute history from the invention of writing, encountered similar problems. Though the origin of civilization was not so metaphysically challenging an issue as the creation, it presented the same problem of determining a starting date, for written sources are a sign that human culture has already begun, and the first historian must postdate the origin of civilization. In trying to identify the origins of history La Popelinière adopted deductive arguments rather than empirical ones, arguments which had no need of absolute chronology. Moses, he said, could not have been the author of the first history, for he lived under the Egyptian empire, which must have had both sciences and history, since without them no civilized state was ever renowned.[9]

La Popelinière shared with Bodin a sense of the importance of chronology to history. He said that history had more need of the knowledge of time than of all the other graces which antiquity had ascribed to it. But lacking a chronological system that was truly universal, he could not give chronology priority over historical narrative. "I would not believe that history is more aided by chronology than vice versa, for the ancient historians are well known and read, while the ancient chronologists have disappeared except for their names" (p. 53).

His dissatisfaction with the ancient chronologists is easy to understand. None of them used a method that could have satisfied his needs. They did not seek the level of integration among disparate cultures which he considered essential to the construction of a complete history. The chronologies of Eusebius and Bede were relative, capable of integrating events already connected in the chronologer's mind and designed to synchronize specific processes. They could not bring together the widely separated cultures that fascinated the sixteenth century. For that century an absolute chronology was necessary. It could not be limited to the calendars and epochs traditionally included but must extend to all possible human cultures.

To satisfy this need for his contemporaries Joseph Scaliger created the first major new chronological system in the West since Bede's. Scaliger (1540–1609) was part of the tradition of French scholarship. His father, Julius Caesar Scaliger, produced scholarly works on grammar and engaged in public dispute with Erasmus. Joseph studied under the French jurist Cujas and edited classical authors, including Catullus, Tibulus, and Propertius, before turning to chronology. From these early scholarly endeavors he brought the standards of humanist scholarship to his chronological work. He decided to produce a chronological system which would stand up to the demanding criteria that he had applied to classical texts.

The fame his earlier success brought gave him great confidence in his chronological system, which he felt was complete and unconditional. It could be used to compute with perfect accuracy the chronological relationship between any cultures and could locate any series of events precisely. In the preface to his major chronological work, the *Opus novum de emendatione temporum*, he wrote, "From our arguments we see thus how a completely absolute chronology [chronicon absolutissimum] can be made,"[10] and he continued throughout his work to express himself with similar assurance.

To fashion a system that would be complete and self-contained Scaliger divorced chronology from moral and religious functions and made it a purely numerical discipline. The number produced by his system indicated only where in time a specific event or series of events occurred, not what it meant or what its moral significance was. Reversing a long tradition that used chronology to establish the date of the last judgment and to verify the accuracy of biblical prophecies, Scaliger looked upon his system as only a tool for dating, without any regard for its religious significance. He accepted the value of nonbiblical sources and stressed the technical difficulties of his discipline. Where earlier chronologers saw their discipline as a guide to students in reading the Bible, Scaliger proposed his study as a field for advanced scholars, who could follow the complexity of his arguments and calculations.[11]

There are two basic elements to Scaliger's method: a new standard of accuracy and a method for treating each year as a unique point. To achieve the first, he sought to establish as accurately as possible the precise years and divisions of former calendars. In pursuit of this goal he exploited the astronomical data that had accumulated over the sixteenth century. In addition he drew on an unusually large number of scholarly studies, including those of Near Eastern sources as well as of classical ones. Finally, he analyzed and classified these data more carefully and thoroughly than his predecessors. He insisted that part of the problem with previous chronology lay in ignorance of the dates found in existing sources. His effort to clarify the calendars occupied the first section of the *De emendatione*.

But lack of data was only one difficulty; more important was the inability to identify each year as a separate point and distinguish it from any other year by intrinsic characteristics that were unrelated to external events. Scaliger felt his predecessors had failed to create a universal chronology because "They did not apply a 'character' and notation to the year which they had conceived of." [12] Because their work lacked this character, earlier chronologers could not incorporate their information into a dating system that intrinsically comprehended all the events of human history. As he finished his review of ancient dating systems and turned to the chronology of the whole world, Scaliger offered a means of creating such a character.

Scaliger's goal had important affinities with Dionysius Exiguus' in setting up the Easter table. Dionysius wanted to find a means to identify each year so that the date of Easter could be calculated. He found that if he knew the phase of the moon and the day of the week—i.e., if he could fix the date in the lunar and solar calendars—he could identify the year within a cycle of 532 years (the 19-year lunar cycle times the 28-year solar cycle). This 532-year cycle was more than enough for the Easter calendar; in fact 95 years proved sufficient for any practical use. But 532 years was not long enough for Scaliger to establish a universal system which would identify all years in which the events of absolute history had occurred. For that he had to find a cycle which would be long enough to include all human history from creation through his own day and would extend for a considerable distance into the future.

To create such a cycle Scaliger started with the 532-year cycle of Dionysius, but he saw that another one must be added to it. "Inasmuch as we need to designate each epoch with a character, the order of times [ratio temporum] will require not only the lunar and solar cycles but also the indictions" (bk. 5, Preface). The indiction was a civil cycle of fifteen years, at the end of which a census was to be taken for tax purposes. Diocletian had established the common indiction at the end of the third century, though it had been in use before then. The census was never taken after Diocletian's time, but the period of the indiction was a standard means of dating charters and other legal documents throughout the Middle Ages. The indictions themselves, like the Dionysian cycles, were not counted. The phrase "the third indiction" meant the third year of the fifteen-year cycle, not the third fifteen-year period after the Diocletian census. A historian trying to use the indiction as a means of dating a charter would need additional information, such as the name of the reigning monarch, or a reference that would permit location in the solar cycle, such as the day of the week.

The indiction allowed Scaliger to create a cycle sufficiently long for his purposes. By multiplying the Dionysian 532-year cycle by the fifteen-year cycle of the indiction he produced a cycle of 7,980 years, which he called the Julian period, since the years it counted were the years of the Julian calendar (so-called after Julius Caesar, who introduced the solar year of

365 and one-quarter days to Roman dating in 45 B.C.). Since the three smaller cycles that made up the Julian period were all commonly used as real dates, Scaliger simply needed to find an event which was dated in all three cycles to establish the location of the Julian period in reference to all other calendars. Many such events were available, but he chose the birth of Christ, an event not only important in itself but one that had been used to date events of the Christian commonwealth.

To calculate the birth of Christ in the Julian period, Scaliger used the work of Dionysius Exiguus, to which he added his own reckoning of the indiction. Dionysius had established the conventional date for the birth of Christ as the first year of the lunar cycle, the ninth of the solar cycle, and the third of the indiction. To find the location in the Julian period Scaliger began by establishing in what year in the 532-year Dionysian cycle Christ had been born. He was not born in year 1, since that was the first year of the solar cycle; similarly he was not born in year 20, which is the first year of the second lunar cycle, since that would be the twentieth year of the first solar cycle; nor was he born in year 39, the first year of the third lunar cycle, since that would be the eleventh year of the second solar cycle. The year which is the first year of the lunar cycle and the ninth of the solar cycle is 457. (Arithmetically, 457 is the smallest number which is divisible by 19 if 1 is subtracted and also by 28 if 9 is subtracted.)

Now Scaliger had to factor in the year of the indiction. The year 457 is the seventh year of the indiction (457 divided by 15 leaves a remainder of 7. The quotient is irrelevant since the indictions are not numbered consecutively.) To bring the remainder around to 3, the year of the indiction in which Dionysius put the birth of Christ, Scaliger would have to add 11, so he needed to find a number which, multiplied by 532 and divided by 15, would leave 11 as a remainder. The answer is 8; 8 times 532 equals 4,256, plus 457 from the original cycle equals 4,713, which is the year of Christ's birth in the Julian period.

The result was especially convenient, since the 4,713 years before the birth of Christ were enough to include any possible calculation of the creation from the Hebrew calendar. By multiplying the three cycles Scaliger had created a chronology that would comprehend all the events of human and divine history and would run almost 1,700 years into the future. With this instrument he could integrate all the civil and religious calendars he had collected and studied, could correlate all previous dating systems, and could locate any event or series of events completely and unambiguously on a single time line. He had devised an absolute dating system whose numbers were independent from any specific series of events. The system was admittedly artificial, bringing together things that had no intrinsic connection and putting them on a time line that none of his sources had used. But Scaliger defended the artificiality. "The intervals of time are like the intervals of space; just as spatial measurement can be combined, so can temporal measurement" (bk. 5, Prologue).

In the remaining books of the *De emendatione* Scaliger dated major events from the creation to Constantine, using for each event the traditional dating within its own epoch and calendar, as well as the date according to the Julian period. In the last books he assembled tables for integrating all known calendars and epochs. These tables are without events, pure collections of numbers. They comprehend fourteen headings, including the years from the Argonauts' expedition, the years of the world, the epochs of Nabonassar, the Seleucids, the years from the birth of Christ. All fourteen are divided into intervals that allow Scaliger to synchronize them with the Julian years. The contrast with Eusebius is striking. Where Eusebius sought to bring together the spiritual and secular sequences, Scaliger's intention was to synchronize all events with one another regardless of their meaning and to produce a single time line.

The arithmetical precision of the table, however, did not stand on its own. It required support from Scaliger's source criticism. The numbers had no intrinsic application to real events unless scholarly method was applied to the ancient sources to determine the accuracy with which events had been dated in their traditional time frames. Scaliger used for this purpose a clear and reliable method drawn from the scholarly principles of his own day and dependent on the notion of absolute history. He had applied this method in his own scholarship on classical texts and was sure it would work with chronology. He felt that the most trustworthy source was always that which was closest in time to the events it described. The earliest source was the most reliable.

His reliance on the earliest source led Scaliger into difficulties, and his attempt to resolve these difficulties shows that he did not assume an absolute time line for his absolute history. In 1606 he published a second chronological work, the *Thesaurus temporum*, a collection of new chronological sources together with another manual of chronology, the *Isagogici canones*. In it he sought, among other things, to reconstruct the Greek chronicle of Eusebius, of which only a Latin version was available in the West. As part of this effort, he included some excerpts of Eusebius that he found in a Byzantine world chronicle (Grafton, p. 170).

One of these passages was a list of the dynasties of Egypt purporting to come from Manetho, an Egyptian priest from the third century B.C. Scaliger's method convinced him that these lists were more accurate than those in Herodotus' history. Though Herodotus was earlier, he had not read the lists himself but had learned their contents from the priests he talked to during his visit to Egypt. Manetho had taken them directly from the most ancient sources, and, being an Egyptian, could read them himself. If Manetho's account was valid, then the length of the dynasties posed major problems to Scaliger's Julian period, for they preceded the creation by 1,336 years. Thus Scaliger's scholarly method led him to accept a source which began the dynasties of Egypt not only before creation but before the Julian period itself.

To resolve this difficulty and maintain the universality of his chronology Scaliger placed another cycle of 7,980 years before the original Julian period. "We postulate that one Julian Period be assumed and called 'the first Julian Period of proleptic time' or the postulated Julian Period." [13] The solution is an obvious one, given the arithmetical origins of the Julian period, and Scaliger justified his act by claiming to be following in the footsteps of the mathematicians. [14] To a modern reader the addition of another cycle seems to present few problems in itself. The real difficulty is the idea of events happening before creation, but that is potentially solvable by rearranging events to produce an earlier date for creation or by assuming the simultaneity of some of the dynasties on Manetho's list. Later scholars tried both solutions, and in future years, when absolute time had come to be widely accepted and incorporated into the European worldview, the idea of an earlier Julian period was not shocking and could be seen as simply an arithmetical convenience. In the comprehensive chronological manual *L'art de verifier les dates* creation is dated in 7731 of the "period anticipating the Julian period," with the Julian period beginning in the 251st year of the world. [15]

But for Scaliger the problem was not a simple arithmetical one, involving only the manipulation of numbers. The Julian period, because of the way it was constructed, was the complete and absolute period of historic time. It embraced the absolute history of Bodin and La Popelinière. The 7,980-year cycle was not an arbitrary number, however artificial it might be; it was the product of natural and civil time. The numbers could synchronize various calendars because they had a real relation to those calendars and included all possible dates.

Thus Scaliger had to explain the nature of the time produced by postulating an earlier Julian period. Invoking the authority of the classics, he argued,

> Varro established three different times: hidden, mythic, and historical. We tend to equate the first two, even though they differ. For many times are shown to be hidden, which are not mythic;

and many mythic events happened in historical time. But we call all that interval proleptic and enroll on the list of the mythic those things which were done before the epoch determined by Moses, as is fitting for a Christian. That time the Hebrews called the age of the confusion of things, in which time there was nothing in the nature of things. Which usage the Greeks call Oxymoron, for something was done when there was no time. (*Isagogici*, p. 273)

This passage is striking both for Scaliger's deliberate association of his time with earlier times and for the perseverance of multiple times in his thought. The Julian period continued a trend toward ever more embracing and comprehensive chronologies that began with the Hellenistic period. Scaliger's chronology included all history and permitted the notion of absolute history to be implemented. His chronology thus recognized the historical reality that dominated historians of his own time. Just as Eratosthenes devised a chronology to record the integration of cultures in the Oecumene and Bede developed one for the implementation of personal time, so Scaliger devised a dating system for absolute history. Like his predecessors, his chronology did not look beyond the subject of integration to establish an absolutely comprehensive time frame. Events falling outside the chronology needed another dating system, expressing another time.

Scaliger was certainly more aware than his predecessors of the difficulties into which he had fallen. Events lying outside absolute history seemed to challenge the very concept his Julian period was designed to implement. His attempts to justify his hypothesis of proleptic time are strained and unconvincing to modern readers. Mythic events are either embellishments of actual historical events or they never happened at all—and thus should be made subject to the techniques of historical scholarship elaborated in the sixteenth century. In either case it is hard to see how they can happen in non-time. The literary term oxymoron can explain the usage, but it leaves the reality as much in the dark as it ever was.

Thus in Scaliger's system multiple times persevered within a method that claimed to integrate all historical epochs and calendars. The chronological table included with the *Isagogici canones* shows the confusion dramatically. The epochs are listed in chronological order, but under three headings. First, under the "Postulated Julian Period" we find the first four dynasties from Manetho's list, the beginning of the Julian period of the Eastern church, and the birth of Adam according to the Septuagint Bible; second, under the "Current Julian Period" we find the fifth to the ninth dynasties from Manetho's list; finally under the "Epochs of Historical Time" we find the creation and all subsequent historical epochs. The Julian period itself is dated according to the Egyptian calendar, in the thirty-third year of the reign of Mencheres in the fourth dynasty (p. 123). Thus the times not only coexist but they overlap one another and are used to date each others' beginnings and endings.

DOMENICUS PETAVIUS AND THE B.C./A.D. SYSTEM

Three distinct interests dominated the development of Western chronology down to the seventeenth century. From Hellenistic times, astronomy and mathematics had served to create systems that approached time as matter in motion. From early Christian times, theology had sought a system which would portray the linear history of humanity as a coherent and unified expression of the will of a personal God. By the Renaissance, the complex and contradictory quality of existing systems had led to the development of chronology as a branch of classical scholarship. These three interests were all present in the work of sixteenth-century thinkers but had not coalesced to produce a unified approach to chronology. In the life and career of Domenicus Petavius they came together in a system that expressed the absolute time appearing among scientists and philosophers during the seventeenth century.

Petavius, born in Orléans in 1583, had completed a degree in philosophy at Paris when he met a Jesuit who stimulated his religious vocation and subsequent entry into the order. After a course of theological studies, he began a career in teaching and writing that produced a number of religious treatises, including works on the Holy Spirit and on the importance of frequent communion.[16] These theological interests were only part of his career, for he also taught rhetoric and acquired the mastery of classics and scholarly technique demanded by that role. In the midst of his theological studies he prepared a series of editions of classical and Christian writers, including Sinesius, Themistius, Nicephorus of Constantinople, and most important of all, Saint Epiphanus, whose chronological work he found especially important. He accompanied all these editions with scholarly annotations in which he demonstrated a growing concern for the problem of dating.

In addition to these scholarly and theological interests, Petavius developed a keen taste for astronomy and mastered this important aspect of seventeenth-century learning to an astonishing degree. After a period in which he taught himself, he asked a noted astronomer to tutor him. During the third lesson, the astronomer stalked from the room accusing Petavius of mocking him, since it had become clear that the pupil knew as much as the tutor (Vital Chatelain, p. 204).

Petavius did not see these interests as unconnected or separate parts of his learning. He felt a great need to integrate them into a single view of life, to make use of science to reinforce his faith, and to bring his scholarly technique to the service of scientific questions. This synthetic approach characterized his attitude from his early career. Petavius' first lecture as professor of theology at Paris was on the necessity of using the profane sciences to interpret the Scriptures. The Jesuits had long insisted on integrating the various disciplines, a tradition that went back to Renaissance education with its stress on the general consciousness that lies beyond all studies. Petavius brought these interests to bear on the field of time in order to make some key distinctions that would enable him to assess the similarities and differences between time periods with great precision and effectiveness.

Scaliger had ended the *De emendatione* by dating the year of its publication in each of the various calendars studied in that work, showing his interest in synchronization without committing himself to one time line. In striking contrast to these multiple sequences and times Domenicus Petavius began his major study of chronology, the *Opus de doctrina temporum*, by asserting the unity of his subject. "Time is one most holy and sacred thing. . . . Just as God contains all, so time is spread out over all things; it is absent from no one, present to all; it measures and changes all things; it will prescribe the spaces in which everyone lives, acts, or endures." [17]

Fundamental to Petavius' method was a sharp division between chronology and history. Though the division was commonplace in Western thought, it was so important to Petavius that it determined the very organization and structure of his works. The *Opus de doctrina temporum* was published in two volumes, the first a technical study of astronomical rules and civil cycles, and the second a practical application of this technical chronology to historical problems. This division seemed strange to his readers, and in a later work, the *Rationarium temporum*, he complained that his first book had been poorly received because of it. [18] Though he blended the two subjects together more fully in that second work, he was careful to distinguish them conceptually and made clear he was doing it only in the interests of readability.

He separated history and chronology the better to show the priority of the latter. In *De doctrina* he noted that history was blind without the order it got from chronology, quoting with approval the classical proverb, "Chronology is the eye of history" (pt. 2, Prolegomena, 2.iii). It is in the Preface to the *Rationarium temporum* that he discussed the matter most fully. There he was nervous about mixing historical and chronological material together and felt it important to distinguish their significance clearly.

> It is thus a thing of intermediate sort that we are beginning, so we should put both [history and chronology] in order so that each expresses also that to which it is chiefly committed. History has as its own to possess fully the matter of deeds and to write down their order, usually with proofs, arguments, and witnesses, whence the order of individual years is established. Chronology indeed inquires after one thing, by what signs and marks each thing may be arranged in its years and times, and is nearly always content with that. It does not extend further than individual events. (Preface)

Chronology then has no part to play in judgments on meaning or in organizing events into a larger whole. It belongs not with the historical disciplines but with the natural sciences, and it is one of the four sciences which have to do with time. In *De doctrina temporum* he mentioned the different questions about time that belong to physics, astronomy and music. Then he claimed there was still a fourth group of questions concerning what kind of civil divisions human beings have used to measure time, "since all peoples civilized and barbarian divide up time into civil parts" (pt. 1, Prolegomena, 1.xxv). Because there is a pure science of time, the application of computation to historical events is a separate field from pure computation. The act of establishing a date should keep the different principles clearly in mind in the interests of precision and accuracy (pt. 2, Prolegomena, 2.iv).

Since the pure calculation of time is conceptually separate from its application, measurement is distinct from judgment in much the same way that Descartes, writing ten years after the *De doctrina temporum*, would separate them in his *Discourse on Method*. Petavius' chronological system thus contained attitudes which historians and scientists in the century after him would find congenial. He created a separate science of time which looked upon it as a continuous, universal reality that could be measured by computational technique. By applying this measurement to individual events, the historian could place them in an absolute time line that was already established and that had an objectivity beyond the reach of earlier historians.

To achieve this result Petavius had to lay down rules for chronological precision. "If there is any art to chronology, it must have rules and principles by which it can establish an order" (pt. 2, Prolegomena, 2.iv). He identified three basic principles that had to be applied to any date. In order to be true a date must be established by demonstration, by hypothesis, and by authority. Though this tripartite division seems to echo Bede's analysis of the parts of chronology into nature, convention, and authority, there is a decisive difference. For Bede the three categories explained the origins of three separate chronological units. The twelve-month year is natural in that it depends on the sun and moon; the months are by custom divided into thirty days; and the week has seven days by scriptural authority. For Petavius the three categories are not historical explanations of separate units but analytical techniques applied to a unified time. They produce a certainty of chronological orientation which transcends the limits of custom and authority.

After laying down these rules and principles, Petavius gave his readers an example of how they could be applied to establish the absolute date of an event in the distant past: the beginning of the Peloponnesian War. To establish the date by authority, he analyzed the sources on which it was based. Abandoning Scaliger's simple reliance on the most ancient, he applied a variety of tests, focusing on the contemporaneity of historians to the event, the agreement of historians with one another, and their reliability in recounting events that we know to be true from other sources. He brought to bear all of the critical techniques developed by scholars in the previous century. In the case of the Peloponnesian War Petavius found two reliable sources. Thucydides was a contemporary of the events, as he told us at the beginning of his work, and Xenophon was a contemporary of later events of the war. Both have reputations for accuracy, and since their temporal references are consistent with one another, we can accept their own dates as authoritative.

Once he had established the dates by authority, he could make them absolute by abstracting them from the temporal processes in which they were embedded and attaching them to an independent temporal sequence. This required two steps: demonstration and hypothesis. Demonstration involves the mathematical analysis of astronomical evidence and it produces results as certain and irrefutable as any mathematical proof. Critical to such demonstration are the eclipses, for they join the solar and lunar cycles at a fixed point. Since all calendars are either solar, lunar, or a mixture of the two, the eclipses provide a point of reference allowing synchronization of any particular calendar with the Julian calendar. In the case of the Peloponnesian War Thucydides mentioned an eclipse after midday in the first year of the war, and Xenophon mentioned a solar eclipse in the twenty-eighth year. By astronomical calculation Petavius could identify the month in which the war began.

Finally, to establish the absolute date, Petavius applied the third operation: hypothesis. He required a date to which by convention all other dates could be referred. He considered the conventional date for the birth of Christ suitable because so many subsequent historians used it. Even though the date is established by convention, the dates which depend on it are necessary and unquestionably valid. "That I am writing in 1627 is true not by demonstration but by convention, but it still cannot be disproved. It must be accepted by hypothesis" (pt. 2, Prolegomena, 2.iv). By calculating from the Olympiad and using the character or the unique number that locates the war in the Julian period, Petavius established the date of the beginning of the war as 431 B.C. This was the absolute and unequivocal date for the event; it depended on no particular historical perspective and was true for all observers.

The conventional point of reference, the birth of Christ, had itself no temporal location. It represented for Petavius not the actual event but an agreed upon point from which all real events could be dated. In the *Rationarium temporum* he located the actual birth separately from the Dionysian date, explaining that it was not necessary that events be dated from the actual year of Christ's birth. What was essential was that the conventional date be fixed and isolated from all other events (pt. 2, bk. 1, chap. 9). Petavius used that date entirely out of convenience without any of the moral associations that Dionysius Exiguus had ascribed to it when he devised the system, noting simply that the Latins numbered from the nativity, as the Greeks numbered from the Olympiad, and others from the creation of the world (2.1.4).

Petavius' system implied a single, continuous, and uniform time. Its certainty and reliability stemmed from the mathematical analysis of authority to locate events on such a time line. He could not accept Scaliger's notion of proleptic time, a mythic time before historical time. No authority, however ancient or believable, could gain acceptance if it demanded such a hypothesis. No one, he insisted, who claimed to talk about the history and chronology of real times ["chronologiam et verorum temporum historiam"] could postulate years or could place the origin of empires before the flood or creation (*De doctrina temporum*, pt. 2, bk. 9., chap. 15). Calling Scaliger's work on the Egyptian dynasty a tissue of lies, Petavius acknowledged the uncertainty of the origins of Egypt, but he felt that Egypt probably began about the same time as Assyria. He denied the authority that Scaliger had given to Manetho and preferred a more balanced comparison of the sources, all of which contained some inaccuracies (2.10.18).

Petavius' attack on Scaliger's dating of the Egyptian dynasties was part of a general polemic against the earlier scholar that dominated De doctrina temporum. Most sections of the first part of the work concluded by pointing out the error in Scaliger's treatment of the subject under consideration. Hardouin, the editor of the 1705 edition, felt constrained to discuss the intense animosity of Petavius to Scaliger and to acknowledge Scaliger's great contributions to chronology. Petavius himself was not so generous. Though he admitted the usefulness of the Julian period and of Scaliger's industry in collecting such a great variety of sources, he looked on his rival's work as part of an old and outdated tradition. He even denied him the credit for inventing the Julian period, noting that the Byzantine historians had long determined the year from the creation by using the indiction together with the solar and lunar cycles. Scaliger had simply applied this Greek method to Roman dates (1.7.7). Petulant though this claim might be, in a deeper sense Petavius was right. Scaliger brought to his chronology a sense of time that still admitted the possibility of a plurality of times in which historical events could contain their own time sequences. Though his Julian period allowed him to synchronize all past epochs and opened the way to an absolute dating system, Scaliger himself did not have a conception of absolute time. Petavius could claim with justice to have brought a new conception of time to his work and to have founded his chronology on a single, continuous, and linear time frame, including all possible historical events but having itself no temporal location. His B.C./A.D. dating system was fit for the use of those who accepted the notion of absolute time.

ISAAC NEWTON AND THE LIMITS OF ABSOLUTE TIME

The assumptions underlying absolute time are mathematical and philosophical in nature, but religion was never far from the intentions of chronologists during the seventeenth and eighteenth centuries. The religious divisions of the period gave to the quarrel between the Catholic Petavius and the Protestant Scaliger the color of a partisan dispute. Considerations of creed did not prevent Protestants from acknowledging the superiority of Petavius' method. As the Dutch scholar Vossius reluctantly observed, "Anyone who devotes himself without partisanship to write about times will discover that though Scaliger deserves great praise, he will prefer to follow Petavius." [19] Vossius himself sought to save Scaliger's reputation by making it appear that he really believed the early Egyptian dynasties were collateral and that proleptic time was thus unnecessary. Alexander Morus called him "the first among chronologers . . . to be followed in preference to Scaliger." [20]

There were issues here that went beyond pure partisanship to strike at the very core of religious truth. Scaliger's use of nonbiblical sources raised doubts about the antiquity of the Kingdom of Israel and its precedence over the pagan empires. As scholars came to see a single continuous time in which the events from all empires occurred, the process of synchronization made the position of Israel seem incongruous to the pious. How could the events of early Hebrew history, before the glorious reigns of David and Solomon, be so significant if they occurred during the time when the mighty empires of Egypt and Assyria dominated the world? More especially, how could the spiritual and intellectual contributions of the Hebrews depend so absolutely on the choice and favor of God, when the Greeks had founded a culture so many centuries before?

To answer such questions Isaac Newton spent forty years of his life constructing a new chronology, published as the *Chronology of Ancient Kingdoms Amended* in 1728. There he proposed a fundamentally new chronology, one which saw Israel under the reign of Solomon as the first great empire, antedating the great dynasties of Egypt, the culture of the Greeks, and the fall of Troy. This conclusion was as incredible to scholars of his own day as it is to us, and to reach it Newton made use of the chronological system of Petavius, all of whose major works he owned. [21] He applied his own conviction of absolute time to historical problems and used the system with even more rigor than his Jesuit predecessor had.

Newton connected the various historical times of the world into a single historical series to show that there was only one time to serve as the basis of his calculations. All time was ultimately traceable to astronomy. He felt that the growth of navigation under the earliest Egyptian kings was the impetus to the study of the stars that led to the fixing of the solar year and the subsequent measurement of the equinoxes and solstices. Once discovered in Egypt, the calendar was taken to Babylon during the reign of Nabonassar by Egyptians fleeing the Ethiopian invasion. These exiles "carried thither the Egyptian year of 365 days and the study of astronomy and astrology and founded the aera of Nabonassar." [22] Newton's source for this conclusion was a passage from the Hellenistic scholar Diodorus Siculus, [23] but that scholar only mentioned the study of astrology as a gift of the Egyptians to Babylon. Newton was concerned with a far deeper contribution, and he expanded the scope of his source to include the measurement

of time itself, not simply its astrological interpretation. He expanded
the meaning of his source deliberately, for in other sections he quoted
Diodorus and cited him as evidence for the Babylonians' fame as astrologers.

Absolute time played an essential role in Newton's redating. In all crucial
cases where he adjusted important events, his method rested on assumptions about abstract numbers. He rejected or accepted specific sources on
the ground that mathematical analysis made them improbable. Such an
approach produced eccentric results and major errors, for he tried to use
sources written in relative time to produce absolute dates. In one dramatic
example he shortened the period of Greek history prior to the Persian wars
by lessening the average reign of the kings listed by Herodotus.[24]

> The Greeks had no chronology before the reign of the Persians
> under Cyrus, Cambyses and Darius Hystaspis, but began then to
> collect it from the number of the kings who had reigned in the
> several cities of Greece, recconing their reigns equipollent to gen-
> erations and three generations one with another equipollent to
> 100 years. And particularly they placed the return of the Hera-
> clides 80 years later than the taking of Troy and collected the time
> of that return from the number of kings who had reigned at Sparta
> after that return by putting their reigns one with another at about
> 35 years a piece. Whereas kings reign one with another but 18 or
> 20 years a piece at a medium according to the course of nature.
> And since Chronology has been certain there is scarce an in-
> stance of ten kings reigning in continual succession above 250
> years. Whence all the times of the Greeks preceding the Persian
> empire are to be shortened in proportion of about 19 to 33 or 4
> to 7 and by this recconing the Argonautic expedition and the
> taking of Troy and return of the Heraclides will be where we have
> placed them.

Newton failed to grasp the limitations of his source. Herodotus did not
conceive generational dating in numerical terms, and numerical conclusions based on his generations are anachronistic and intrinsically erroneous.

Even more striking in their use of absolute time are Newton's astronomical proofs, which he regarded as "the surest arguments for determining things past."[25] He used astronomical analysis to redate two major
events, in both cases erroneously. He moved the Argonautic expedition—
and with it the fall of Troy—from the traditional date in the twelfth century B.C. to the middle of the tenth; and he removed 11,000 years from
the period of the pharaohs to place the reign of Amenophis in the period
just after the voyage of the Argonauts. The effect of his redating was to

place the beginnings of Egyptian and Greek culture after the flowering of Hebrew society in the reigns of David and Solomon.

To redate the Argonautic expedition Newton assembled several bits of astronomical evidence from fragments in existing sources dating back to the Hellenistic period.[26] On the basis of this evidence he felt he had reconstructed the actual astronomical sphere made for the Argonauts. Called the Primitive Sphere, since it was the earliest example of its kind, this was intended to serve as a navigational guide in the Argonauts' expedition. Newton knew that Hipparchus, the Hellenistic source, had misjudged the precession of the equinoxes. Since Hipparchus' source for the sphere, the fourth-century astronomer Eudoxus, did not know about the precession at all, Newton reasoned that Eudoxus had located the equinoctial and solstitial points not as he saw them in the fourth-century sky but as he found them in the original source. Thus Hipparchus found them as they were located in the sky when the Primitive Sphere was drawn up for the Argonauts, and he modified the points according to his mistaken assumptions about the precession. Newton recalculated the points according to the correct measure of the precession and derived the date of the original sphere.

The extraordinary feature about Newton's conclusion is the sharp contrast between the numerical certainty of his date (he dated the sphere in 939 B.C.) and the tenuousness of the evidence upon which some of his most crucial calculations were based. He had no assurance that Eudoxus had copied the original sphere exactly—or even that it was Eudoxus who passed the sphere on to the Hellenistic astronomers. He certainly could not be sure that the original sphere was accurate. He was aware of the limitations of his evidence, noting at the end of his discussion, "Tho these descriptions are coarse yet by their help we may come pretty neare the truth."[27] Newton's doubts were not strong enough to moderate the precision of the numbers, however, which in turn reinforced his confidence that he had redated the Argonautic expedition.

> From all these circumstances, grounded upon the coarse observations of ancient astronomers we may reckon it certain, that the Argonautic expedition was not earlier than the reign of Solomon, and if these astronomical arguments be added to the former arguments, taken from the mean length of the reigns of kings, according to the course of nature; from them all we may safely conclude, that the Argonautic expedition was after the death of Solomon and most probably that it was about 43 years after it. (*Chronology*, p. 75)

The process by which Newton made these calculations is not unlike that he used to calculate the precession itself and to reconcile lunar and terrestrial motion. Later editions of the *Principia*, without significant new data, manipulated existing numbers to produce greater precision. This precision depended on assumptions that treated the complex geology of the earth in mathematical terms to explain the relation of the tides to the precession and to correct the miscalculations of the first edition.[28] Like the precepts of natural science, Newton's chronology depended on the assumption of abstract, numerical motion and of continuous time and space that possessed an inherent order and precision. He felt he could dismiss the imprecision of the data in history as easily as the discrepancies in the data of natural science. They were only a means to an end. The numerical order of events in time depended for Newton not on the concrete data but on the structure and definition of time itself.

By recalculating the astronomy of the Primitive Sphere Newton had made Greek culture younger than the flowering of the Hebrew kingdom, but there remained the problem of Egypt, mentioned in the Bible itself as a great empire during the period before the Hebrews even entered Palestine. Newton felt that the changes in Greek dating could lead to a reassessment of the antiquity of Egypt. "Let us now try to rectify the chronology of Egypt, by comparing the affairs of Egypt with the synchronizing affairs of the Greeks and Hebrews" (p. 142). To change that chronology he dismissed the earliest dynasties, including the pharaoh of the Exodus, as mere local rulers who could not be called founders or rulers of empires. Even eliminating these early rulers did not solve the problem, for Amenophis, and his world-conquering son Ramses, antedated the reign of Solomon by centuries in the traditional dating scheme.

Faced with this challenge, Newton used astronomical data coupled with important assumptions about his sources to move the reign of Amenophis down to the period after the Argonautic expedition. His key piece of evidence was Amenophis' tomb. Like the Primitive Sphere, it no longer existed. Cambyses had defaced it during the Persian conquest, but Diodorus Siculus had found an account in Hecataeus which said that Amenophis had inscribed on his tomb a circle divided into 365 parts for the days of the year. For each day the makers had given the time of the rising and setting of the sun. This description convinced Newton that the times were calculated in order to establish when the vernal equinox occurred and to give the year an astronomical beginning instead of the traditional beginning marked by the rise of the Nile. He further reasoned that since the circle was part of the memorial to Amenophis, the calculation of the astronomical beginning of the year occurred first during that pharaoh's reign, with his death marking the first year in the series.

If Amenophis was the first to date the year from the vernal equinox and if that year was indeed taken to Chaldea by Egyptian exiles, then Newton could establish a precise date for the reign of Amenophis. The Egyptian year of 365 days fell almost one-fourth of a day behind, each year. The first year of the era of Nabonassar fell 33 days and five hours before the equinox, and it would take 137 years for such a discrepancy to accumulate. Thus Amenophis died around 884 B.C., or 96 years after the death of Solomon (pp. 60–62).

Just as he had redated the Argonautic expedition, Newton introduced here, under the guise of astronomical precision, highly questionable assumptions that are not supported by his historical data. Not only was his source for the inscription on Amenophis' tomb indirect, but there were no grounds for assuming that the discrepancies in the Chaldean calendar were directly related to the Egyptian calendar. He was even less justified in assuming that Amenophis was the originator of the Egyptian solar year. The chain of reasoning in his argument was entirely linked by his commitment to absolute and continuous time. That commitment dominated the selection and interpretation of his evidence and gave him license to read conclusions into it that were unwarranted from the point of view of existing scholarly standards.

Absolute time is related to absolute space, and Newton used the spatial aspects of the Egyptian kingdoms to reinforce his chronology. His redating forced a reconsideration of the temple begun by Amenophis and continued through the reign of Psammetichus. The Egyptian chronicles ascribed 11,000 years to its completion, introducing several dynasties between Amenophis' son Ramses and his successor. Arguing that no building could last that long, Newton felt that the temple had been built in two to three hundred years. He analyzed the building with his preconceptions of continuous space, just as he had determined the date of Amenophis' death with his preconceptions of continuous time. His treatment of the temple reflects his general conception of history. As Frank Manuel observed,

Newton's chronological writings might be called the mathematical principles of the consolidation of empires because they dealt primarily with quantities of geographic space in a temporal sequence; the individuals mentioned in his histories, usually royal personages, were merely signposts marking the progressive expansion of territories; they have no distinctive qualities. The subject matter of his history was the action of organized political land masses upon one another; crucial events were the fusion of previously isolated smaller units or the destruction of cohesive kingdoms by quantitatively superior forces. (Manuel, *Isaac Newton*, p. 137)

So important were these astronomical arguments to Newton that he used them even when they were confusing and unnecessary. Discussing the period in which Hesiod lived, he said, "Hesiod tells us that sixty days after the winter Solstice the Star Arcturus rose just at sunset: and thence it follows that Hesiod flourished about an hundred years after the death of Solomon, or in the generation or age next after the Trojan war, as Hesiod himself declares" (p. 75). The reference to Arcturus is obscure and resists interpretation to this day (Manuel, p. 65). More important, even if it were clear it would add little to other evidence. Newton had evidence from Herodotus and Hesiod himself to date his life. In the short chronicle which begins the *Chronology* Newton omitted all reference to Arcturus and explored more fully the relevant evidence from historical sources. In his chronological analysis, however, he obviously felt that astronomical evidence must be used wherever it could be found. Even if it did not allow us to fix a date more precisely, such evidence provided an important connection with the astronomical world that represented absolute time most closely and that most nearly approximated a sense of time as pure number.

VICO AND THE CRITIQUE OF ABSOLUTE DATING

In Newton's work the mathematical and numerical aspects of absolute time had run riot, producing results that few chronologers were willing to accept. Those who had some association with Petavius were particularly incensed. Hardouin, the eighteenth-century editor of the *De doctrina temporum*, launched a fierce attack on Newton's method, noting the absence of firm textual evidence for his historical assumptions and challenging his conclusions from beginning to end (Manuel, chap. 10). He cited evidence from classical literature to show that no one had ever considered Chiron, the legendary figure to whom the Primitive Sphere was attributed, to be an astronomer. But even this attack is based on the assumptions about continuous and absolute time that are found in Newton's work, for Hardouin used the B.C. dating system to construct a continuous time line in which to establish that Newton's own source was an untrustworthy guide to astronomy.[29]

Hardouin, who had been so solicitous to acknowledge Scalige·'s strengths, was vitriolic in his attack on Newton. The Jesuit Hardouin was not simply attacking the Protestant enemy; he was alarmed at Newton's misapplication of principles implicit in the chronological method he thought so reliable. Nor was Petavius himself without the tendencies that wreaked such havoc with Newton's chronology. A modern chronologist has observed that Petavius, in synchronizing the eras, had a tendency to assume an order where in fact there were a series of coincidences.[30]

Newton was not without supporters. Voltaire, who instinctively supported anyone the Jesuits disliked, regarded Newton with intense admiration and accepted his dating. In his *Letters concerning the English Nation*, he summarized the key points of Newton's chronology accurately and sympathetically. Others, especially among the English scholars, were similarly approving. But by the nineteenth century few could be found to place the beginnings of Greek and Egyptian culture after the reign of Solomon.

The dating system of Petavius nevertheless survived the excesses of Newton. His work became the touchstone for anyone seeking to synchronize the events of the ancient past and to construct a universal chronology. Practicing historians like Gibbon were aware of his work, though few studied it in the original.[31] Chronological treatises came to assume the superiority of his dating system for the synchronization of all possible human cultures. In *L'Art de vérifier les dates*, published in 1819, his work is fundamental, though the editor acknowledges key ambiguities in the method (1:xxvi). In particular, astronomers and historians differ on the use of the year zero. Historians do not use it as a date; astronomers do. Thus scholars applying astronomical data to historical events risk producing a discrepancy of one year.

Against the chorus of acceptance there was an important dissenting voice. Giovanni Vico, one of the most original thinkers of the eighteenth century, felt that the chronologists of the sixteenth and seventeenth centuries had erred in the focus on astronomical time.[32]

> Those two marvelous geniuses, Joseph Justus Scaliger and Denis Petau [Petavius], with their stupendous erudition, the former in his *De emendatione temporum* and the latter in his *De doctrina temporum*, failed to begin their doctrine at the beginning of their subject matter. For they began with the astronomical year, which, as noted above, was unheard of among the nations for a thousand years, and in any case could have assured them only of conjunc-

tions and oppositions of constellations and planets in the heavens, and not of any of the things that had happened here on earth
nor of their sequence. . . . And on this account their work has
shed little light on the beginnings or on the continuation of universal history.

Why was Vico so concerned about the astronomical dating system? His
displeasure stemmed from his disagreement with one of the prevailing notions of seventeenth and eighteenth century science. Though he admired
Newton and sent him a copy of the *Scienza nuova*, Vico held a basically
different idea of truth. Where Newton and Descartes felt we could acquire
absolute truth about nature through mathematical analysis, Vico thought
that we could never know the truth about nature. Only the creator could
know the truth about his creation. Since God created nature, only he
could know natural truth. What Descartes and Newton mistook for truth
was certainty, a feature growing out of the mathematical method. But the
certainty of mathematical conclusions was entirely self-contained; it did
not imply a true understanding of nature.

Human beings could only know the truth about what humanity had created. "The rule and criterion of truth is to have made it."[33] We could know
the truth about society, politics, and culture, since they were human creations, but we could know the truth about mathematics or geometry only
to the extent that we saw them as culture, not when we required conclusions based on them to have a necessary relationship with a natural world
outside human creativity. History was the study of what we have created,
and through history we would know the truth.

With this assertion Vico not only challenged the scientific world of his
own day but set for himself the task of developing a method for understanding history as a whole. In the *Scienza nuova*, published in 1725 and
substantially revised in 1731, he set out to establish such a method. Since
human creation makes history truthful, a reconstruction of human history
as a whole must depend on myth, poetry, and all other human creations.
Where Newton had euhemeristically considered gods and mythical figures
to be based on historical individuals, Vico countered that myths expressed
human consciousness at a particular stage of development and revealed
deep truths of the human past. To understand human history we must
understand the mythic content of those first times when the basic structures of language and human interaction took shape.

Historians must start then at the beginning, treating history as a coherent product of human creativity. In the *Scienza nuova* Vico dated
events from the beginning of the world and considered the accounts of the

earliest periods of human history, found especially in the Scriptures, to be the most important that we have.

> Sacred history is more ancient than all the most ancient profane histories that have come down to us, for it narrates in great detail and över a period of more than eight hundred years the state of nature under the patriarchs; that is, the state of the families, out of which, by general agreement of political theorists, the peoples and cities later arose. Of this family state profane history has told us nothing or little and that little quite confused. (*New Science*, par. 165)

Vico, like Newton, constructed his method by combining theoretical assumptions with existing sources, but the focus was quite different. Where Newton used assumptions about continuous and absolute time to fill gaps in his sources and create specific dates, Vico used contemporary assumptions about the origins of society to point up key weaknesses in current historical method. Political theorists of the eighteenth century, such as Montesquieu, had traced civil society to the family, and if these theorists were right, then we must understand a period of time when society was familial, for which we have only sacred or mythic sources.

Vico's use of Petavius' chronology in this passage is also interesting. The eight hundred years of the Patriarchs is a period consistent with the *De doctrina*, where the period is dated from 2328 to 1521 B.C., and Vico's editor feels that Petavius may be the source for the figure. Vico, however, avoided any suggestion of absolute dates and looked at the period as part of a particular process within human history, given meaning and significance not by its absolute location but by its own status as the prepolitical, familial society. In other sections of his work Vico maintained this pattern of using chronologists' conclusions but rendering their dates into his own relative time frame. In one of the few passages where he made direct reference to Petavius, he mentioned the Heraclides and the Cretan dynasty of the Curetes. Omitting the carefully calculated absolute date, he wrote, "These two great fragments of antiquity, as Denis Petau observes, fall in Greek history before the heroic time of the Greeks" (77).

History for Vico was the history of human thought and consciousness, and it must start at the beginning, with mythic thought. "Our science is therefore a history of human ideas, on which it seems the metaphysics of the human mind must proceed. This queen of sciences, by the axiom that 'the sciences must begin where their subject matter began' took its start when the first men began to think humanly and not when philosophers began to reflect on human ideas" (347). Here was the crux of his attack on the chronological tradition of his day. Petavius did not start at the beginning, for by the time people had acquired astronomical dating many of the crucial events of history had already occurred. To recreate these crucial events he drew on mythic and scriptural sources and invoked an overall pattern which he felt inhered in human history itself, put there by the inner structure of the human consciousness that created it. This pattern was a three-part cycle from the age of Gods to that of heroes to that of men. The human stage, though the culmination of the process, is not stable and soon breaks down, causing the cycle, or *corso*, to recur.[34]

The cycle made it possible for us to understand the stages of prehistory, those earliest times before documentation. By looking at human history in its earliest documented stage, we can make judgments on the processes that had preceded it. Our judgments in turn can be confirmed by using the myths these prehistoric ages had passed on.

> Thus purely by understanding, without benefit of memory, which has nothing to go on where facts are not supplied by the senses, we seem to have filled in the beginnings of universal history both in ancient Egypt and, in the East, . . . the beginnings of the Assyrian monarchy. For hitherto this monarchy, for lack of its many and varied antecedent causes, which must have been previously at work in order to produce a monarchy, which is the last of the three forms of civil government, has appeared in history as a sudden birth, as a frog is born of a summer shower. (738)

Vico's method of reconstruction is dramatically different from Newton's. Both used preconceptions to fill a gap in the historical record, but Newton assumed a natural truth based on the existence of continuous and unchanging space and time. Vico assumed a historical truth based on human creativity with a beginning that determined its subsequent structure. Vico's knowledge carried its own necessity. We cannot tolerate ignorance of our beginnings, and they must be reconstructed before we can have any awareness of our true selves.

It was precisely the continuous and unchanging nature of the chronologists' time that bothered Vico. At one point in his summary of world history he came to a succession of ten kings. "These royal successions are great canons of chronology. . . . According to the rule of the chronologers," their reigns took up 300 years. Vico objected strongly to the principle of calculating a period of time based on numerical assumptions about the length of generations. "But Thucydides says that in heroic times kings drove one another off the throne almost daily" (75–76). The length of reign for Vico is a particular fact, applying only to a concrete historical circumstance. Later on he came back to this theme, citing Tacitus as well as Thucydides to document the instability of early monarchs and the impossibility of applying to their reigns any numeric rule (645). For Vico time did not exist independently from historical events, and numerical rules that assumed such an existence produced a distorted picture of the past.

The uniformities of the past are intrinsic to the events, and we can discover them only by considering how events were measured when they occurred. To find the beginnings of human history we must recover the two ancient measurements of time that preceded astronomical dating: the measurement of time through the harvest and through astrology.

> In this way chronology has certainty lent to its successive periods by the progress of customs and deeds through which the human race must have marched. For, by an axiom above stated, chronology has here begun her doctrine where her subject matter began: that is, with Chronos or Saturn (after whom time was called chronos among the Greeks), the reckoner of the years by the harvests; with Urania, watcher of the skies for the purpose of taking the auspices; and with Zoroaster, contemplator of the stars in order to give his oracles from the paths of falling stars. . . . Later, when Saturn ascended to the seventh sphere, Urania became the contemplator of the stars and planets, and the Chaldeans, with the advantage of their immense open plains, became astronomers and astrologers. . . . Thence finally mathematics descended to measure the earth, the measurements of which could not be ascertained save by the already demonstrated measurements of the heavens. (739)

In his vision of different times, each with its own measure but all forming part of human experience as a whole, Vico looked forwards and backwards. He harkened back to the historians who recorded human history in the centuries before Newton, measuring time relative to the events being described and using dates that took their structure and import from the meaning of a particular sequence. He looked forward to the time of the twentieth century with his insistence on history as a product of human creativity and on language as the key to understanding truth.

Vico could not present in a single statement the pattern he saw in history. Linear progress was part of the story of humanity in that the movement from Gods to heroes to men was a progressive one in which the highest stage of freedom and the exercise of human will characterized the last stage. Furthermore human history as a whole was progressive because subsequent repetitions of this cycle depended on language and myths formed earlier. The cycle was not perfectly repetitive.

But history for Vico was more a process than a linear progression. Events emerge from underlying structures much as they do in the process of organism which constitutes the basis of Whitehead's metaphysics, and human will itself lacks the focus and unity that had grown out of the Augustinian tradition in the West. The decision making process seems closer to the world of Proust and Freud than to that of Newton and Descartes. Vico rejected the prevailing trend that dominated chronology and history in the centuries after Herodotus. Where each new chronology encompassed wider and wider aspects of human experience into a linear time frame, culminating in the absolute history of the sixteenth century, Vico pointed to historical realities that had been excluded. He denounced the narrowness of the absolute time frame and its tendency to exclude vital areas of human experience. Like Pompeius Trogus he sought a synthesis and meaning in his subject that contemporary chronology did not afford. And like Braudel or Foucault he looked to underlying structures of discourse to provide that meaning. In doing so he discovered the need for a new sense of historical truth and a new means of verifying assertions about the past. But his method was not new; it simply went against the prevailing method of source criticism that applied to events in absolute time. The sense of historical truth that inheres in relative time differs considerably from the natural truth that is inextricably linked to absolute time. The narratives that measured time in relative terms needed a different system of validation from the one in common use since Ranke. They needed a system which would not exclude some of those aspects of human experience which were essential to a full understanding of its meaning. The following chapter will explore the issue of truth as it applied to the histories written in relative time. It will show the dimensions of reality that came into play as different perspectives on truth grew out of different historical subjects.

8

Conclusion: The Truths of Relative Time

As to the speeches [logoi] that were made by different men, either when they were about to begin the war or when they were already engaged therein, it has been difficult to recall with strict accuracy the words actually spoken, both for me as regards that which I myself heard, and for those who from various other sources have brought me reports. Therefore the speeches are given in the language in which, as it seemed to me, the several speakers would express, on the subjects under consideration, the sentiments most befitting the occasion, though at the same time I have adhered as closely as possible to the general sense of what was actually [alēthōs] said. But as to the facts of the occurrences [erga] of the war, I have thought it my duty to give them, not as ascertained from any chance informant nor as seemed to me probable, but only after investigating with the greatest possible accuracy [akribeia] each detail, in the case both of the events in which I myself participated and of those regarding which I got my information from others. And the endeavor to ascertain these facts was a laborious task, because those who were eye-witnesses of the several events did not give the same reports about the same things, but reports varying according to their championship of one side or the other, or according to their recollection.

Thucydides, *History of the Peloponnesian War*, 1.22

Though the affinities between history and literature are strong, historians are not simply poets. They describe things that actually happened. Com-

mitment to factual accuracy is the key difference between them and novelists, poets, or dramatists. Almost without exception there has been broad
agreement on this issue since the first historians. Aristotle's characteriza-
tion of the difference between history and poetry was among the earliest
statements of this position, and it has served as the touchstone of subsequent comments. "The distinction between historian and poet is not in
the one writing prose and the other verse—you might put the work of
Herodotus into verse, and it would still be a species of history; it consists
really in this, that the one describes the thing that has been and the other
a kind of thing that might be." [1]

Aristotle, as a natural and social scientist in search of general laws and
universal truths, felt that this distinction left history in a subordinate position. "Hence poetry is something more philosophic and of graver import
than history, since its statements are of the nature rather of universals,
whereas those of history are singulars." Historians, of course, have not
been so ready to accept Aristotle's judgment. Polybius, for example, emphasized the usefulness of his subject:

> The tragic poet should thrill and charm his audience for the mo-
> ment by the verisimilitude of the words he puts into his charac-
> ters' mouths, but it is the task of the historian to instruct and con-
> vince for all time serious students by the truth of the facts and the
> speeches he narrates, since in the one case it is the probable that
> takes precedence, even if it be untrue, the purpose being to
> create illusion in spectators, in the other it is the truth, the pur-
> pose being to confer benefit on learners. (2.56)

However people have evaluated the relation between history and truth,
they have generally agreed that the connection is indissoluble. Except in
rare cases of incorrigible partisanship or extreme incompetence historians
have told the truth as they saw it, have avoided knowingly presenting a
fable as a fact, and have embraced the accuracy of their discourse as the
root and foundation of history's very usefulness. Without the connection
with actual events, history loses an essential part of its function and value
to the reader. On this key point all of the historians discussed here agree
with the most exacting of modern historians. Ranke and Thucydides speak
with one voice.

But the assumptions about time that separate us from the historical tradition before Newton carry with them distinctive definitions of truth and
accuracy. These earlier historians discriminated the true from the false,
the event from the fable, in ways that differ significantly from ours. In part

the differences are technical and reflect modern developments in scholarly and critical technique, but the basic difference is conceptual and springs from the nature of relative time. Absolute time sees all events as potential points on a single time line. The events that form the basis of historical judgments and generalizations can be located on that time line with certainty and simplicity. The possibility of simple location provides a standard of accuracy and precision for all other historical generalizations. At its root, a historical account consists of particulars whose temporal and spatial location is indisputable. Even though the time or place of some event might be in fact unclear due to the absence of documents, in theory all historical events, given appropriate evidence, can be located without ambiguity.

In relative time, where not all events can be simply located, such certainty is not available. Though events might be established beyond dispute in one time frame, they exist in others where they might take on a different order. Their significance shapes the temporal relations among them, and generalizations about that significance are an essential part of the process of validation. The strategies for establishing the truth in this time are necessarily different from those that obtain in the world of absolute time. For historians who used these strategies, there were still events whose visibility or intellectual clarity made it possible to locate them unequivocally on a given time line. But there were also events which were not so clear, which existed in an episodic time frame or in more than one linear series. Such events might be important in determining the significance of the story as a whole and thus might play a role in establishing the proper time line on which to place the datable events. But they themselves could not be validated by simple location.

From Herodotus to Guicciardini there were two distinct means of validation. One was appropriate to events that could be precisely and quantitatively located on a given time line, and another was suitable for events and historical realities that could not. The first means of validation sought to establish the veracity of events as precisely as possible. The second cannot be defined so simply. In one sense the truth sought was that of verisimilitude, and the events so validated were only presumed to be probable, more likely to have happened than any other, given the available evidence and the realities of human nature. But verisimilitude does not adequately comprise all of the elements of this second form of historical validation, for it also represented a shift in focus. Where validation through veracity concentrated on the relationship between the historians' accounts and the actual events, the second form of validation focused on the historians' relation with the readers, conveying the honesty of the account and the extent to which it represented their best efforts to describe the past without partisanship and bias. Where historians sought veracity, the events could be conceived as independent realities, existing as objective points on a time line; where they used the second means of validation, there were inescapable psychological and moral factors that created an intrinsic bond between the historian and the reader and that transcended the objective material existence of any particular historical fact.

Thucydides' method, as he expressed it in the passage quoted at the beginning of this chapter, depended on these two means of validation. In his statement he used two different words for truth: "akribeia" or accuracy, and "alētheia" or sincerity and honesty. They combined to produce wisdom, or the *saphes* that Thucydides hoped his readers would derive from reading his account of the war. That wisdom is based on the validity of his interpretation, the accuracy with which the events are recounted, and the overriding similarities in human nature that connect the events of all times with one another. Such wisdom is the goal of all human thought, philosophical and poetic as well as historical. It was in the minds of most Greek thinkers the purpose of humankind and the highest attainment of the human mind. Thucydides reinforced the seriousness of his purpose in his readers' minds by contrasting his efforts with those of more dramatic and sensationalistic writers.

> And it may well be that the absence of the fabulous from my narrative will seem less pleasing to the ear; but whoever shall wish to have a clear view [saphes] both of the events which have happened and of those which will some day, in all human probability, happen again in the same or a similar way—for these to adjudge my history profitable will be enough for me. And, indeed, it has been composed, not as a prize-essay to be heard for the moment, but as a possession for all time [ktema te es aiei]. (1.22)

In this description of his method Thucydides sought to convince his readers that he had presented the events in such a way that this saphes could be acquired. To establish the validity of his account he divided his work into two separate parts, the events (erga) and the speeches (logoi), each with its own standards of verification and each with its own goals. He verified the concrete events of the war by means which we clearly recognize as part of the critical technique of most historians. He sought out eyewitnesses and compared their accounts to eliminate biases and distortions that might arise from the limited perspectives of the viewers. By applying this critical method he felt he had presented the events with accuracy (akribeia).

Akribeia is a word denoting precision and calculation; it applies to events that can be clearly and quantitatively located in time. Thucydides used the word in other parts of his history to describe calculations of time. He accused his predecessor, Hellanicus, of inaccuracy in his chronology (tois khronois ouk akribōs) and defended his own chronology based on years of the war with the claim that reliance on the lists of officials led to inaccuracy.[2] Akribeia is important for all those events that existed in the linear time of the war. We have seen in Chapter 3 how Thucydides conceived these events as part of a single linear series. Their meaning depended on their relationship to one another in that series, and he accepted the necessity of establishing with accuracy the veracity of the events he included.

Thucydides did not claim to have applied these techniques to his whole work. In the case of the speeches he felt that the obstacles to akribeia were overwhelming, and he used another technique for them, one designed to produce alētheia rather than akribeia. This antithesis between speeches and events seems somewhat artificial to us. To the extent that a speech was actually delivered, the difficulties of reporting it do not seem different in kind from those of reporting a battle or a revolution. In both cases there will be several witnesses, each seeing or hearing a different part. The task of the historian is to compare the various accounts and sift out biases and inaccuracies. In place of the sharp distinction he made between events and speeches, Thucydides could just as easily have said, "All parts of my subject were hard to recreate accurately, the speeches particularly so, since no one could remember the exact words used by the speakers. Faced with this special problem, I have given the structure of the argument as I heard it or as it was reported to me, comparing other eyewitness accounts to rule out bias and to compensate for faulty memories. The speeches are thus reported as accurately as possible."

To have described his method thus would have blurred the fundamental distinction in his work between events which exist in simple linear sequence and events which do not. He did not seek akribeia for the speeches, even in those cases where it was possible to do so. What he intended was alētheia, a term which carries connotations of honesty and frankness. Precision was inappropriate, for the material in the speeches did not exist in the sort of time for which precision can be acquired or which can be analyzed numerically.

Thucydides' choice of another means of verifying the truth of the speeches has a momentous impact on his work as a whole, for the speeches are integral parts of his narrative. They occupy one-quarter of the work and contain some of the most rigorous and clear analysis of the causes and motives behind the war. The differences in character that underlie the conflict emerge in the speeches. Not only that but some material that modern historians would see as part of the "events" appears only in the speeches. When the Epidamnian revolt precipitated the war, there was a treaty in force between the Athenian Empire and the Lacedaemonian Confederation. Much depended on the terms of this treaty, especially whether Athens violated its terms when it admitted Corcyra into its alliance. Despite its importance Thucydides never described the treaty in the narrative, referring to it only in the Corinthian and Corcyraean speeches. There the Corcyraeans maintained that it permitted any neutral power to join either side, while the Corinthians pointed to the provision which prohibited either side from accepting a new partner for aggressive purposes (1.35,40). The treaty existed for Thucydides not as actual words written in a document and as an event in the linear time of the war but as a perception in the minds of the participants. He also used the speeches rather than the narrative to assess the strengths and weaknesses of the two sides in the war (1.120–24,140–44). He felt that each side's perception of the other's strategic position was a more important factor in explaining the course of the war than the actual quantifiable strengths seen apart from any perception.

The speeches are not, then, rhetorical ornaments. They are not even pure explanatory sections based on the events recorded accurately in the narrative. They are integral parts of his account of the war, containing material without which the events of the narrative would be incomplete and meaningless. They record and analyze events. They are themselves events in that they are the focus of key decisions in the course of the war. But they do not occur in the same time as the events of the narrative, and they cannot be represented with akribeia.

Thucydides was oddly indifferent to whether the reader believed that the speeches were actually delivered in the linear time of the war. Some speeches were certainly delivered, and Thucydides had access to reliable information for most of the speeches he put into his work. But some speeches could not have been given in the way they occur in the work. The author certainly fabricated them—or at least altered them significantly. In Book One the Athenian ambassadors at Sparta replied in detail to Corinthian arguments they could not possibly have heard, since the Corinthians spoke at a closed session of the Lacedaemonian Confederation's council. In describing that council meeting Thucydides emphasized

its secrecy and the desire of those present to keep its details from the Athenians (1.67). He did note that the Athenians had heard about the meeting but gave no hint how they discovered the exact details of the Corinthian arguments (1.72). Had he been genuinely interested in
convincing his readers that the Athenians had actually spoken in the fashion he described, he could easily have suggested some specific source for
their information, a common practice in the narrative sections where he
wanted to verify his facts.[3]

The Melian dialogue in Book Five represents an even more important
example of a fabricated speech. Melos was a small island that remained
neutral throughout most of the war. At the time of the dialogue the Athenians had decided that neutral islands constituted a temptation for her
allies to revolt and insisted that Melos join the Athenian Empire. When
Melos refused, the Athenians besieged the island and subjugated it. The
dialogue purports to be a discussion between the Athenian and Melian
representatives just before the final Athenian attack. In it the Melians
raised issues of morality and justice, while the Athenians replied that the
powerful always did what they wished to the weak. "For of the gods we
hold the belief, and of men we know, that by a necessity of their nature
wherever they have power they always rule" (5.105).

The Melian dialogue is an important part of Thucydides' overall interpretation of the war, for it raises some of the deepest concerns which led
him to become its historian. The right of the stronger, which stimulated
Athenian aggression towards Melos, wreaked a terrible vengeance on the
Athenians themselves as they invaded Sicily, where they did not have the
strength to conquer and where they were annihilated in a climactic scene
of Thucydides' work. He placed the dialogue in the narrative just before
the decision to go to Sicily, and many of the arguments of the Melians are
those raised in vain by the opponents of the Sicilian expedition.

Despite its thematic importance, the dialogue almost certainly did not
take place in the form in which Thucydides presented it. Both the form
and the content speak against its veracity. In formal terms the dialogue is
not a series of paired formal orations but a more natural conversation with
short questions and spontaneous answers. Statesmen and generals did not
use this form in actual diplomatic practice, since the Greeks held them
incompetent at this kind of interchange and restricted their diplomatic activity to formal speeches.[4] In the matter of content the Athenians had real
and substantive complaints about the actions of the Melians as well as
some grounds for questioning their neutrality.[5] Athens' allies supported
these complaints, and they would certainly have been the subject of any
real discussion between Athens and Melos. Had Thucydides wished to

give akribeia to the dialogue he could easily have done so by alluding to the specific grievances of the Athenians and noting the reasons why they adopted this unconventional form of debate. He did not do so because he wished to endow the account not with akribeia but with alētheia.

Alētheia differs significantly from akribeia in the relationship it implies among the historian, his audience, and the past. Akribeia focuses on the relationship between the account and an objective reality outside the narrator. Alētheia by contrast focuses directly on the intentions and the sincerity of the narrator and does not speak to the relation between the judgment and the concrete events. That is not to say that the judgment—or any of the other contents of the speeches—was subjective. The objectivity, however, lay in the relationship between the author and reader. The historian's analysis, descriptions of motives and character, moral judgments, all depended for their effectiveness on the general sense among the readers of how decisions were made, what values applied in what circumstances, and how society functioned. Alētheia thus became a process of ascribing verisimilitude to events, of showing that they were like truth. For the events to which alētheia was ascribed, akribeia would be inappropriate, since the topics of the speeches had no precise and serial temporal relations either to each other or to the concrete events which made up the linear time of the war. The Melians themselves replied, when constrained by the Athenians to confine their remarks to practical rather than moral considerations,

> As we think, at any rate, it is expedient (for we are constrained to speak of expediency, since you have in this fashion, ignoring the principle of justice, suggested that we speak of what is advantageous) that you should not rule out the principle of the common good, but that for him who is at the time in peril what is equitable should also be just, and though one has not entirely proved his point [entos tou akribous] he should still derive some benefit therefrom. (5.90)

The distinction between akribeia and alētheia, between precision and an honest approximation to the general truth, is hard for modern thinkers to grasp. We tend to feel that precision should be the basis of all objective judgments and that approximation of verisimilitude should be based only on those facts that can be established with certainty and precision. But for Thucydides, as well as the two millennia of historians that came after him, these were parallel approaches to the truth, necessitated by the differing qualities of the times that made up the past of human experience. To say that Thucydides did not mean the speeches to be accurate is anachronistic.[6] He meant them to be as accurate as his account of the events of the war but not in the same way. He understood that to demand akribeia of all aspects of the war would be to exclude its most useful and important dimensions from his readers' consideration.

Alētheia, the historian's honest attempt to represent a truth that transcends quantification and numerical precision, is basic to the speeches, but it intrudes even into the narrative when more than one time frame is necessary to a full understanding of the events. In the case of the naval battle between Syracuse and Athens discussed in Chapter 3, Thucydides placed the battle in two linear times, one the linear time of the war and the other the linear time in which the Greeks had learned their mastery of naval technique. In order to place the battle in the second sequence, Thucydides had to ignore the absence of Syracusans from the earlier battle and create a linear sequence where strict akribeia would not have established one. He could do so because his judgment was an honest attempt to understand the two processes. The connection between the two battles existed honestly in the mind of the historian, a key factor in the establishment of alētheia, and allowed him to enhance the usefulness of his work and to show the progressive qualities of this temporal sequence of learned naval tactics.

Because alētheia does not conform precisely to Newtonian concepts of objective truth, we might be tempted to regard it with suspicion and to doubt its truth value, but classical historians did not regard alētheia as an opportunity to introduce fantasy and exaggeration into their narratives. Polybius, who was even more determined to avoid easy entertainment and sensationalism than Thucydides, defended forcefully and unambiguously the truth of his narrative. He criticizes one of his sources, Phylarchus, for the overly dramatic descriptions of lamentation,

> Leaving aside the ignoble and womanish character of such a treatment of his subject, let us consider how far it is proper or serviceable to history. A historical author should not try to thrill his readers by such exaggerated pictures, nor should he, like a tragic poet, try to imagine the probable utterances of his characters or reckon up all the consequences probably incidental to the occurrences with which he deals, but simply record what really [alētheian] happened and what really was said, however commonplace. (2.56)

Like Thucydides Polybius saw alētheia as an appropriate goal for both erga and logoi, and also like his much admired model, he did not feel that alētheia demanded the precise adherence to the literal truth that akribeia demanded. When he recorded a speech made by the young Philip of Macedon before his council, he observed,

> Finally the king spoke, if indeed we are to suppose that the opinion he delivered was his own; for it is scarcely probable that a boy of seventeen should be able to decide about such grave matters. It is, however, the duty of us writers to attribute to the supreme ruler the expression of opinion which prevailed at his council, while it is open for the reader to suspect that such decisions and the arguments on which they rest are due to his associates and especially to those closest to his person. (4.24)

Veracity and verisimilitude overlapped and existed simultaneously in Thucydides' account of the war. They were each essential to the lasting value of his work, its claim to be "possession for all time." They coexisted easily because Thucydides, like all other classical historians, was more interested in the typical than the unique. As he made clear in his statement of method, he felt his work should be useful forever because the sort of events he described would recur. When classical historians did encounter events that seemed out of the ordinary, their first impulse was to find the general rules that made them part of the overall pattern of reality. Polybius, though he acknowledged a role for chance in history, felt that historians should be reluctant to accept it as an explanation for unusual events. "We must rather seek for a cause, for every event whether probable or improbable must have some cause" (2.38). History needed to bring out the similarities and resemblances among events, and to do so the rhetorical ascription of similar values, motives, and political realities was as important a tool as the science of enumeration with its claims to exactness and precision.

The central concern of Christian historiography threatened to dissolve this symbiosis between veracity and verisimilitude. The uniqueness of the incarnation placed that event beyond the reach of arguments which aimed only at verisimilitude. The resurrection of Jesus in particular is an event completely outside the probable. It depends on the testimony of witnesses and demands accuracy and precision, not simply the honest judgment of the historian as to its meaning, probability, and larger significance. Verisimilitude did not thereby disappear from the arguments of Christian historians. They merely shifted its focus from the event to the source. Since the testimony of witnesses was the basic evidence for the key events of human history, Christian historians sought means to evaluate the credibility of their sources. Among these means were many of the arguments from probability that Thucydides applied to his witnesses, including their access to the event, their possible bias and limitation in reporting it, and the agreement among several witnesses. The difference is that once the credibility and reliability of witnesses were established, their evidence was accepted even though it contravened conventional standards of probability and testified to facts that could not be made part of a general view of how things happen.

Concern with the source focused on two issues among medieval historians, the intentions and the intelligence of the witness. To validate a source, the historian tried to show both that the events were reported honestly and to a good purpose—that the reporter's intentions were good—and that the witness knew how to interpret what had happened correctly—that the intelligence was sufficient to evaluate the experience. These two concerns are clearly present in Augustine's *Confessions*, which are suffused with his desire to validate the story. Since no one could document or cross-check the most important events, especially the scene in the garden, the story depends on the reader's confidence in Augustine's honesty and intelligence. To bolster confidence in his intellect, he showed his skill at using arguments from verisimilitude to interpret those aspects of his life which were common to all his readers. In describing his infancy, for example, he cited behavior that could be seen in most infants and used the notion of original sin to give this behavior a new and convincing meaning. He gave the reader confidence in his honesty by his willingness to accuse himself of sins. He readily admitted the pride and ambition which clouded his motives in the past and which continued to plague him even at the time he wrote the *Confessions*.

These twin standards of intelligence and honesty, intellect and will, are conspicuous in the scene in the garden. Confronted with the voice of the child, Augustine tried to think if there were a children's game which might involve the words he heard. By this response Augustine showed his reader that the voice appeared to him as a physical one and that he applied standard tests of probability to distinguish it from ordinary experience. These critical judgments help convince the reader that Augustine understood the scene. To persuade us he reported it honestly, he introduced the scene by describing in the frankest terms his depression, psychological instability, and moral weakness. Exaltation of himself or his virtues seems to play no part of his reason for including the scene. The reference to his emotions and his attempts at analysis give verisimilitude not to the scene itself, but to his witness of it and his capacity to represent it accurately to the reader. The scene itself remains unique and never enters the field of probability. The central event, the flooding of light into his soul as he read the words of St. Paul, is a personal experience for which we must trust the author.

Augustine was his own source for the *Confessions*, but the techniques of validation he used were also relevant to the distant past. As their fundamental method of verification Christian historians tried to establish a chain of honest and competent testimony leading from the event to the historian. Bede began his own history by describing Albinus, his principal source, as both reverent and learned, taught by other learned and respected men (*History of the English Church*, Preface). Bede specified where Albinus got his own information, tracing the line back to the events themselves. "The story of this miracle [concerning Bishop Aidan] is no groundless fable; for it was related to me by Cynimund, a most faithful priest of our own church, who had it from the mouth of the Priest Utta, on and through whom the miracle was performed" (3.15). Bede, like Augustine, cited the intention of the source not only to reinforce the character of the witness but to demonstrate the function of the account. "In this convent [Barking] many proofs of holiness were effected, which many people have recorded from the testimony of eyewitnesses in order that the memory of them might edify future generations" (4.7).

As they described the distant past, the most serious issue Christian historians of the Middle Ages faced was the reliability of the Scriptures. The events recorded in the Bible were ultimately to be taken on faith, but even then the problems of interpretation and of textual inconsistencies forced historians to apply some techniques of source criticism. For example, the Septuagint differs from the Hebrew Bible in the ages it ascribes to the Patriarchs. For each, at the birth of his first son, the Septuagint gives an age that is one hundred years more than the age given in the Hebrew Bible. In trying to reconcile this discrepancy, Augustine addressed the intentions of the writers. First, of course, he had to identify where the error had occurred, so that he knew whose intentions he was trying to analyze. Did it occur in the original Scripture or was it the result of scribal error? Even in answering this question he addressed the possible intentions of those involved. He argued that it was unlikely to be the result of conspiracy in either version. It was unlikely that either the Jews, spread all over the world, could have conspired to falsify the Hebrew version so consistently or that the seventy renowned scholars of the Septuagint would have had any reason for falsification. Therefore, Augustine concluded, the discrepancy must arise from a single scribal error, made in the first copy and repeated in all subsequent ones.

By this chain of reasoning, based on analysis of the intentions of the Septuagint scholars, Augustine traced the error to a single individual.

Now he needed to analyze the intentions of this person. The error could not be simply accidental, since it occurred repeatedly in the text. The scribe must have had a reason for adding the 100 years. The most likely one was that he accepted the theory that explained the great age of the patriarchs by the shortness of the year used to measure time. If the years were shortened enough to make the ages of the patriarchs seem normal, then 100 years have to be added to the age at which they first became fathers or they would appear to have sired children before puberty. Thus the scribe's intentions were good in that he wanted to make the Scriptures believable, and he sought to change what he considered to be minor details in order to increase acceptance of the Scriptures as a whole; but he lacked the intelligence to evaluate correctly a false theory of interpretation (*City of God*, 15.13).

The scribal error, traced to a unique person, can be found by looking at the intentions of the author, but the general issue of the patriarchs' ages lies in that area of reality subject to the standards of verisimilitude. In those cases Augustine felt it appropriate to ignore the literal words in search of more general truth. Augustine consequently accepted the biblical estimate of the patriarchs' ages by arguing the general principle that men lived longer in ancient times. To prove this proposition he adduced both the testimony of accepted scientists like Pliny, and his own experience. He himself had seen giant teeth on the shore at Utica. If men could be larger in size in early times, certainly they could have lived longer as well (15.9).

Augustine here separated the problem of the patriarchs' ages into two issues. The first depended on understanding the intentions of a specific individual, while the second involved the general order of nature, where the intelligent analysis of experience—Augustine's own and that of other scholars—was germane. The contrast with Thucydides is striking, for the Greek historian was interested only in the latter issue. In the first book of the *Peloponnesian War* he faced in Homer a source of early Greek history from which he wanted to derive a picture of the events of the Trojan War different from that which tradition had passed down. The post-Homeric legend maintained that the Greeks had followed Agamemnon to Troy out of a sense of honor and because of the oaths of loyalty they had sworn. Thucydides disagreed. "And it was, as I think, because Agamemnon surpassed in power the princes of his time that he was able to assemble his fleet, and not so much because Helen's suitors, whom he led, were bound by oath to Tyndareus" (1.9).

To prove this point Thucydides extrapolated information from Homer but ignored the poet's intentions. "And he [Homer] says, in the account of the delivery of the sceptre, that Agamemnon 'ruled over many islands and all Argos.' Now, if he had not had something of a fleet, he could not, as he lived on the mainland, have been lord of any islands except those on the coast, and these would not be 'many'" (1.9). Thucydides based this judgment on his own experience of the Athenian naval empire in which he was an admiral. From this experience he reconstructed the probable power relations of the Homeric period, applying the preconceptions about human motives and political realities that inform his whole work. He could easily have addressed Homer's intentions in constructing the work the way he did, noting Homer's stress on honor and showing how that interpretation had formed the poem, just as Augustine showed that the scribe's acceptance of the short-year theory led to the 100-year error. But Thucydides did not share Augustine's interest in his source's intentions. His interest was purely in alētheia, and he was satisfied to construct an honest appraisal of the underlying realities of power and motives. The actual reliability and intentions of the witnesses and the extent to which they could be trusted to produce precise accounts (akribeia) did not apply to this aspect of his narrative.

Since medieval historians never completely incorporated the personal time of Augustine into their accounts of political matters, their works display a complex and ambiguous relationship of veracity and verisimilitude. Arguments drawn from classical historians to give probability to political events exist side by side with those seeking to explore the intentions and honesty of sources. Otto of Freising, whose writings combine the Augustinian and classicizing tendencies in medieval historiography, at times used arguments that are reminiscent of Thucydides'. Citing Suetonius' statement that Varus waged war in Germany for three years with fifteen legions, he inferred "how great was the strength of the aforesaid tribe of Germans, since at the time of the greatest power of the Roman Empire it wrought such great havoc with the Roman army" (*Two Cities*, 3.3). Here, in much the same way Thucydides had used Homer, he analyzed Suetonius to form conclusions which were not in the original author's mind. Otto seemed indifferent to the Roman historian's intentions or probity and concentrated instead on extrapolating a probable conclusion about German power from his words.

Despite these classical affinities, the core of Otto's source criticism is Augustinian. At the outset he described his method as one which preferred honesty to sophistication and intelligence.

Nor can anyone rightfully accuse me of falsehood in matters which—compared with the customs of the present time—appear incredible, since I have recorded nothing save what I found in the writings of trustworthy men [probatorum virorum] and then only a few instances out of many. For I should never hold the view that these men are to be held in contempt if certain of them have preserved in their writings the apostolic simplicity, for as overshrewd subtlety sometimes kindles error, so a devout rusticity is ever the friend of truth. (*Two Cities*, Preface)

Otto reconciled contradictions among his sources by looking at their intentions. When he investigated the history of Saxony he found contradictions between his principal source, Frutolf's chronicle, and other accounts of the period.[7] In particular, the different chroniclers offered conflicting interpretations of the legitimacy of Henry of Saxony's claim to independence from imperial suzerainty. Otto resolved these conflicts not by investigating the actual political events and using general principles to reconstruct the political context of Henry's claim, but by psychological assumptions that enabled him to reconstruct the intentions of the sources. "These conflicting accounts of historians resulted, I think, from the fact that since men's intellectual abilities had begun to grow and to keep pace with the glory of empire, as a result, when the imperial authority was transferred to the Franks and men's sympathies [animi] were divided upon the physical division of the kingdom, the writers extolled each his own state as much as he could with the aid of his transcendent abilities" (6.18). In identifying the conflict as a lack of probity on both sides, Otto felt not only that he could safely establish the truth between the two interpretations but that his version would acquire veracity. "I myself, keeping a middle course in these matters, and, so far as I am able and can conjecture from what they have said, holding fast the thread of truth [seriem veritatis], will strive by God's grace to turn aside neither to the right hand nor to the left" (6.18). Though his conclusions were conjectural, Otto described the result as if he had proven the veracity of Henry's relation to the emperor. His conviction of the importance of intentions led him to confuse the two parts of verification, ascribing veracity to an account which by classical standards would have verisimilitude.

In theory, by extending the personal qualities of time to the political elements of the narrative, the humanist historians of the Renaissance should have increased the medieval concern for the honesty and reliability of eyewitnesses. In fact the humanists tended to stress verisimilitude and to neglect available witnesses in favor of arguments from probability. Several factors contributed to this practice. Like medieval historians, they did not fully realize the potentiality of Petrarchan time and wrote narratives in which it defined only a narrow range of events, leaving most political events in the time of the Oecumene. In addition, their heightened interest in classical models led them to imitate classical techniques of verification rather than medieval ones. Furthermore, their concern to apply secular modes of analysis to the understanding of political events also led them to diminish the significance of the faith and trust that are so vital an element in the acceptance of eyewitness testimony to unique events.

The humanist narratives did not omit miraculous and unique events. Miracles occur in Renaissance histories with less frequency than in many medieval narratives, but the humanists were generally more selective in their subject matter, ruling out many nonpolitical areas that had found their way into medieval chronicles. That alone would explain the rarity of miracles. The striking feature of the humanists' treatment is that when they did include miracles, they often ignored eyewitness accounts in order to verify the miracles by arguments from verisimilitude. Where they found in their chronicle sources evidence that the event had actually happened, the humanists omitted that evidence and argued that the event was possible.

A Florentine victory from the early years of the republic affords an illustration of humanist practice. At Campaldino, Florentine Guelphs decisively defeated the Aretines, and, according to Giovanni Villani, a mysterious voice announced the victory to the Signoria in Florence at the very hour it occurred. When Villani recounted this miracle, he validated it in much the same way that Augustine tried to convince his readers that he had actually heard the voice of the child. Villani, himself a member of the Signoria at the time, explained that the Florentines had tried to find a natural source for the voice before accepting it as a genuine apparition. Villani gave his own assurances that he was there and had heard the voice himself, insisting that "this was true. I heard and saw it" (*Cronica*, 7.31).

Bruni authenticated the miracle quite differently. He described the voice but omitted any mention of the priors' attempt to find a source for it. More surprisingly, he did not allude to the presence of reliable eyewitnesses. Villani was his source for much of this period of Florentine history and certainly provided him with the information for this incident. Bruni acknowledged the earlier chronicler's reliability as a witness and historian, but in this case he did not feel that eyewitness accounts were appropriate. The story had for him not veracity but verisimilitude. It was not so much "true" as "like the truth." He established its verisimilitude in two steps. First, to show the abstract possibility that victories might be announced miraculously, he cited classical historians who had described similar announcements in Rome and Macedon. By giving examples of such occurrences in the past Bruni demonstrated the possibility of truth, but verisimilitude requires more than that. He also had to show that this particular announcement had the ring of truth. To accomplish that he concentrated not on the eyewitness accounts of the audible announcement but on the deeper meaning of the intangible qualities of the event. He argued that the growth of Florentine power, of which the victory was an example, was part of a general trend and thus illustrated the divine favor bestowed on Florence. Given this general pattern, "the same divine grace by whose favor victory was won might by the same favor give swift news to those whom He has helped" (4.77).

Bruni's indifference to eyewitness evidence is puzzling to modern readers, but he saw in the incident a reality whose intangible dimensions were more important than its actual concrete occurrence. The announcement existed in that world of psychological states and intentions that underlay the growth of Florentine power and that provided the conceptual structure of his history. The announcement's volitional and abstract existence was more important than its status as a concrete and precisely datable event. Miracles had verisimilitude if they fit into the general conceptual pattern that gave meaning and usefulness to the history as a whole. Villani's concentration on the veracity of the visible event is part of a general tendency on his part. He often allowed his interest in the actual truth of the events, their veracity, to obscure an understanding of their larger significance. When the sudden death of the Lucchese commander Castruccio Castracani removed a serious threat to Florence's independence, Villani filled his story with details of his own experiences and never gave the reader a judgment on the general significance of what had happened. Bruni left out the details but made very clear that a dramatic change in Florence's power had occurred, one he was willing to ascribe to God's work. "And so the danger threatening the city was removed, not so much by human effort as by divine favor" (5.137).

Bruni and Villani disagreed on the role played by the visible, tangible events in history. For Villani, as well as the medieval historical tradition in general, the events, though they might point to deeper levels of being, also existed in their own right. They were all, in a certain sense, figures of the incarnation, which does not simply represent a deeper reality but is itself the ultimate reality. Bruni adopted a more flexible approach to the events. They represented intangible realities, for some of which he could apply secular analysis and for some of which he needed to invoke divine favor. In either case the significance lay in the underlying, intangible substance of the past, not in the visible and precisely datable events. The true reality of the past suffered if the events as a whole lacked verisimilitude, and it was very important that the reader find the interpretation probable; it was less important whether the reader chose to disbelieve some of the visible manifestations of this intangible reality.

The visible events thus became for Bruni only a part of the historian's strategy for establishing the truth of his narrative. He had also to shape them into a meaningful whole, illustrating the analytical and moral values which made the past useful. In the case of Campaldino, even if the reader chose to disbelieve the miraculous announcement of victory, Bruni's account would remind one of the continued growth of Florentine power with its psychological and political roots. That theme does not depend on any particular concrete event but upon the general meaning and pattern Bruni placed on Florentine history as a whole. The events thus became a rhetorical tool, like the speeches, the ascription of motives, and the explanations of military strategy. They were not the ultimate goal of critical method but only one of the steps in establishing the verisimilitude of the narrative.

In their search for verisimilitude the humanists did not mimic their classical models, for they could not recreate the symbiosis that had existed in classical times between veracity and verisimilitude. The thrust of Petrarchan time was against it. The visible, concrete event attained a status that could not be entirely ignored; it could never become purely a rhetorical tool in the interest of usefulness or entertainment. Renaissance historians thus felt a tension between the goals of veracity and verisimilitude that Thucydides or Tacitus did not feel. They feared that the demands of veracity might conflict with history's moral and practical purposes, that the truth of the visible event might diminish the meaning and significance of the past. At the same time they feared that the cause of probability and the demands of rhetoric might rob history of its truth.

This tension in humanist history erupted into a dramatic argument between Lorenzo Valla and the Neapolitan scholar Bartolommeo Fazio, the quarrel of King Martin's snore. The dispute began when Fazio attacked Valla's history of the Neapolitan King Alfonso. Fazio accused Valla of inelegance. Not only was his language full of barbarisms but he included in the work details that were undignified or inappropriate for a royal history. Fazio was particularly incensed by a passage where Valla described the Aragonese King Martin falling asleep and snoring during an ambassadorial audience. Valla used the incident primarily to comment on the king's wit, for when the ambassador stopped his oration on noticing the royal slumber, King Martin asked him to continue, saying, "Though I was snoring with my eyes closed, I was not asleep. Though the body might sleep, the mind does not."[8]

Fazio found the anecdote quite unseemly. "In the first place, what you have said," he wrote to Valla,[9]

> is against verisimilitude and the rules of narration. For a narrative should be not only true [veram] but likely [verisimilem] if you want to be believed. Can it possibly seem credible that a king at an audience for ambassadors should not only fall asleep but even snore? Does this perhaps seem a small thing to you to make him sleep? Or do you not know how indecorous this is for the royal dignity? You seem to want to show us not a king but some drunken man, who has been seized by some torpor or feebleness of the body and was besotted enough to exhibit snoring. Good teacher, you must write to preserve the person's dignity; otherwise your narrative will be improbable and will not inspire faith.

Valla, the target of this attack, has earned the reputation of a true scholar whose critical method formed part of the basis of modern scholarship. Certainly several of his other works reinforce this point of view. In his analysis of the Donation of Constantine he proved that a document in which the emperor supposedly gave the papal states to the popes was actually a ninth-century forgery. He played a major role in the genesis of new legal studies, when he attacked the well-known contemporary lawyer Bartolus of Sassaferrat for ignoring the historical context of the law. His grammatical and rhetorical work established new standards of linguistic scholarship. Even in history he had made contributions by criticizing the reliability of such standard classical sources as Livy.

Valla's reply to Fazio's attack displays some of this concern for factual accuracy, for he did respond to some of Fazio's allegations by asserting the veracity of his narrative. To Fazio's objection that he should not have used evidence from the court jester, Valla replied,[10]

> You blind bat! I am composing a history of the king; now why should a royal history be held in awe more than a royal palace? If that man [i.e. the jester] always had a place in the councils and private chambers of the king, can he not be allowed a little corner in the edifice, as it were, of my history? I might remind you that cooks, horse-boys, and that sort of people are not only present in the dwellings of princes but are necessary members of them; and the same is true in the case of their histories.

He also cited his sources for some occasions where Fazio had challenged his veracity.[11]

But Valla was not a simple defender of veracity against Fazio's insistence on verisimilitude. He agreed with Fazio that history should be like truth rather than the literal fact; he even allowed that the historian might embellish and add events to strengthen the probability of the narrative. Though he argued that the jester had actually had a place in the king's council, in other types of situation he did not make such an argument from veracity. In most cases he responded to the attack by insisting not that what he described had actually happened but that it was more likely to have happened. He argued that the king's snore that had scandalized Fazio gave greater verisimilitude to the work by showing monarchs as they tended to behave in real life. Valla included the snore in response not to a new standard of scholarship but to a different sense of reality, one based on a more realistic analysis of the political situation than that afforded by traditional moral categories. "I have not recorded what people ought to think but what they do think" (1:575–76, in Janik, p. 399).

Valla thought Fazio displayed ignorance in two important areas. First of all he did not know the true place of oratory. "Why will I blame you because in ignorance of the art of oratory you impose the rules of the law courts on me for the writing of history" (1:570)? The law courts might insist on truth, but under the rules that oratory imposes on history, it is more important that a story have verisimilitude than that it be actually true. Invention must not do violence to the truth (veritatis violentiam), but it should embellish the truth to make the reader accept its probability.

Because oratory included the issue of probability, the historian had to know not only the rules of rhetoric but the way human beings actually behave. Here was the second area in which Fazio was deficient. "You know nothing of the human condition [ex rebus humanis]" (1:570). Fazio lacked the medical knowledge to distinguish between sleep and drunkenness, did not understand the types of human behavior which were acceptable in public and those which were not. Certainly it was improbable that the king should appear drunk or lethargic in public, but appearing to fall asleep was quite another matter. Fazio had understood neither the probabilities of royal conduct nor Valla's precise language in describing his behavior.

For Valla there was a crucial difference between the visible events, the erga, such as the court jester's testimony or the actions of King Ferdinand, and the logoi, the words and motives that constituted the second half of Thucydides' method for true and useful history. The court jester's participation was an event as concrete and solid as a palace, but the king's snore had more the features of an oration. It must be treated with the rules of oratory, always keeping in mind that probability and verisimilitude are part of that subject.

King Martin's snore existed in another historical time and space from the court jester's testimony. The fact that the jester commented on the behavior of the court is an indisputable part of the record of concrete events. His evidence needed interpretation, but its temporal existence was clear and uncomplicated. King Martin's snore was ambiguous and not precisely locatable. Valla could not tell if he was pretending. The record could not resolve the issue, for none of the courtiers could be sure. King Martin himself might not have known whether or not he was actually asleep. The snore existed at the margin of consciousness where time is not simply a linear succession but a complex multilayered process.

Valla applied to such historical events as King Martin's snore a standard of verification different from that he used for those which existed in simple linear time. Like Thucydides and the other classical historians, he made them a part of oratory, where verisimilitude was the basic goal and where veracity would do violence to the event. But this attitude is not a license for free invention. Rhetoric needs reality for completeness. The historian must give the events plausibility by applying to them motives and standards of behavior which the reader will find real. His oratorical talents come into play as he tells his stories in such a way that the readers either recognize their own assumptions about human nature or modify and refine those assumptions by confrontation with a more sophisticated point of view. It is precisely in this activity of refinement that history finds its most important use.

Though Valla's history betrayed a division between words and deeds similar to Thucydides', words played a role in Valla's thought that made a clear division between the logoi and the erga impossible. Valla, as much as any humanist, accepted the philosophical implications of Petrarchan time. He felt that rhetoric was superior to philosophy, since the context and usage of the words were the key to the thought they revealed. The events of the past are found in words; we can no more perceive the events directly than we can know abstract truth apart from its expression in a historical context. Thus the questions of intention and discourse, which he placed firmly in the realm of verisimilitude, are crucial to source criticism, to the area of concrete events as well as that of thoughts and motives.

For the Renaissance humanists words bridged the gap between verisimilitude and veracity. Petrarch himself had understood this function, but he did not treat it in such secular terms as Valla. Urging a correspondent to make many friends, the earlier humanist quoted Juvenal on the rarity of good men. Following the quotation he offered several perspectives on its truth.

> "Who said that," you ask? Why do you care? If you approve of the saying, why do you seek the author? All truth, as Augustine says, is true by veracity [a veritate verum est]. I say it. Now do you still deny it? Experience, which is not accustomed to lie, says it; truth, which cannot lie, says it. But if you insist on the mortal author, it was said by Juvenal, an author most skilled in such things, who also knows human customs deeply. If you do not believe him, listen to another, who speaks in the voice of Him who not only knows men but created them. "There is not one good man, no not even one." The poet says few, the prophet says none; and each is true according to his own meaning. (Familiari, 3.15)

Petrarch validated the underlying truth of the words here by a complex combination of experience and authority. He adduced the experience of the author, his own experience, and the absolute authority of the Scriptures, all cooperating to bring experience and authority together into a single expression of the truth.

The special function Valla accorded words is clear in his treatment of the Donation of Constantine. There he was analyzing both a distant historical event and an actual document that still existed in his own day. Insofar as his treatise concerned the distant historical event, he needed to make judgments of verisimilitude concerning the intentions and policy of Emperor Constantine and Pope Sylvester. Insofar as he was writing about the actual document, he needed to make judgments of veracity concerning the precise authorship and date. In Valla's analysis, the problem of verisimilitude occupied the central place, not only taking precedence over arguments from veracity but even providing those arguments with their structure and impact. In his introduction, he told the reader he would prove the document false (falsum), but before he did that "the order required [ordo postulat]" that he first prove that Constantine and Sylvester were not the sort of men the one to give the donation and the other to receive it. To prove that point Valla constructed imaginary orations in which Constantine gave the lands to the church and Sylvester accepted them. As he analyzed these orations, Valla pointed out that in each case the speakers would have to say things that contradicted their deepest principles and ran counter to their behavior at all other points of their careers.

After making these points about the improbability of the actual event, Valla then turned to the document in which the Donation was recorded.

The issue here is one of veracity, for he wished to prove the document a forgery and thus a false record. Nevertheless his treatment of the document makes clear why he consider the order to require a discussion of the issues of verisimilitude first. He approached the document as if it were itself an oration, applying the same standards of probability that he applied in the hypothetical orations of Constantine and Sylvester with which he began his treatise. Coming to the section of the document which bestowed the land of Rome, he pointed out that the document would have to be approved by the Senate. Then he asked the reader to imagine the donation being presented there, asking whether the Senate, so recently converted to Christianity, would accept such a diminution of its rights uncomplainingly. Similar passages recreate imaginary speeches by Constantine's sons and by Sylvester.[12]

Judgments of veracity about the event itself do occur in the treatise. Valla argued that the pope did not in fact take possession after the supposed donation, that a small donation was given to an earlier pope, that existing law would not support the popes' claims. With regard to the document itself he also made arguments based on veracity, showing by philological analysis that the language was commonly used not in the imperial chancery of the fourth century but in the papal chancery of the ninth. He gave examples of events referred to in the document that did not happen in the fourth century. He illustrated the anachronisms and contradictions that fill the supposed donation (1 : 775–76). All of these arguments, however, are framed within an approach which treats both the event and the document as oratorical, and where the issue of verisimilitude is fundamental to acceptance or rejection.

The Donation of Constantine existed in a special sort of time. As a physical thing it was precisely datable, but the intentions and policy behind it were not. By the application of standards of veracity to it, it could be dated in a certain period of time and a certain place, but in order to apply these standards Valla first had to make judgments of verisimilitude to establish probable authorship. The document itself thus maintains an ambiguous status, requiring both standards of verification. Potentially, of course, all sources possess the same ambiguity. They exist in Petrarchan time, where historical context determines meaning. They also exist in a more public time, where judgments of veracity based on unequivocal location in a linear sequence are appropriate and where the ambiguity of intentions has no place. The sources express in their very concreteness the deep tensions of Petrarchan time, tensions it derived from its Augustinian roots, and which a millennium of historians had not reconciled.

Two possible lines of development lead out from Valla's work, each offering to resolve the tension in humanist history between veracity and verisimilitude. If scholars could apply Valla's philological method to all historical evidence, then they could treat the past as a body of concrete events whose veracity could theoretically be established, even if in some particular cases the evidence was too scanty for this to be done in practice. This is the path adopted by sixteenth- and seventeenth-century scholarship. It led to the critical achievements of the French humanists, the complete history of La Popelinière, and the absolute time of Petavius and Newton. It was the foundation of nineteenth-century methodology and the irreducible substance of modern historical research. It is also inconsistent with the assumptions of relative time.

But there is a second possible development from Valla, one which would focus on verisimilitude as the basic standard of historical verification and embrace the proposition that rhetoric and history cannot be disentangled. This approach treated the problem of verification as a rhetorical one, to be resolved by the application of reasonable standards of human conduct and the realistic representation of the past through narrative. Though such a solution seemed overly literary and backward in the nineteenth century, contemporary thought has become increasingly sympathetic to a point of view which treats reality as a system of discourse. This attitude certainly has strong advocats in linguistics, philosophy, and sociology, but in recent years historians too have given it more serious consideration.

Machiavelli and Guicciardini followed this path. Their works, though the result of considerable research and investigation, did not contain elaborate exercises in critical method. Guicciardini, the more diligent in research of the two, made extensive notes on dates and the location of major events, leaving blanks in his manuscript so that he could ascertain the correct date for a subsequent draft, but he did not tell the reader of his effort. Nor did he refer often to his sources, though he had access to the papal records and the Florentine archives and could easily have documented some of the diplomatic and military events of his narrative. He was so indifferent to proving which sources he used and how he used them that Ranke regarded him as an unreliable reporter of sources. Only in this century have historians of the Renaissance rehabilitated his reputation as a diligent and conscientious researcher.[13]

Machiavelli seldom referred to his sources. He mentioned Bruni and Poggio in the introduction to his *Florentine Histories* and admitted his debt to them, even though he criticized their stress on foreign policy. He never engaged in explicit criticism of their accounts, nor did he allude to his own research. He treated his narrative as if it were a restructuring of material that his readers agreed on, even though his account was a judicious blending of contradictory accounts found in the chronicle sources, as well as in the humanist histories.[14] His explicit judgments concern the wisdom or folly of specific decisions or courses of action, not the accuracy with which his sources may have recorded a series of events.

Machiavelli certainly cared whether or not his readers accepted his narrative as true. He felt the lessons to be drawn from the past were crucial to overcoming the crisis his nation was passing through. He sought to convince his readers not by explicit source criticism which could establish the veracity of external events but by rhetorical means that went to the structure of the narrative itself. He brought the reader to see the events not only as visible realities but as expressions of human perception and consciousness. In his work the demands of veracity and verisimilitude were joined into a single narrative gesture which combined the chroniclers' use of eyewitness evidence with the humanists' interest in the general significance of the event within a conceptual framework.

Machiavelli's technique of combining his concrete evidence with a general assessment of significance is clear in his treatment of miraculous events. He omitted the miracle at Campaldino, but he did record a number of extraordinary events, the most famous of which was the destruction of the top of the Duomo by lightning at the time of Lorenzo de' Medici's death. "And as many catastrophes would result from his death, Heaven gave numerous clear signs. Among these the highest point of the Church of Santa Reparata was hit by a thunderbolt with such force that most of the pinnacle collapsed, to everyone's great stupor and wonder" (*History of Florence*, 8.36). By referring to the amazement of the bystanders Machiavelli left no doubt in his readers' minds that there were eyewitnesses to the actual lightning bolt and that it had veracity as an actual, datable event. At the same time he drew his readers' attention to the significance of Lorenzo's death and the impending doom that hung over Italy, giving the event verisimilitude as part of trends that could not be precisely dated.

At another point Machiavelli told of an unusual storm that swept Tuscany in 1456. As with Lorenzo's death, he began by saying that the heavens made war on the earth and that the effects produced by the storm would appear marvelous and unaccountable to posterity. At this juncture, earlier humanists might have given the event verisimilitude by referring to storms in their own experience or in classical literature. Poggio Bracciolini had mentioned an incident that happened while he was in Rome during the pontificate of the antipope John XXIII to give support to his account of the miraculous announcement of Pope Gregory's death (2.76–77). Machiavelli's own chronicle source had stressed the availability of eyewitnesses.[15] He instead described in some detail the storm itself, including exact dates, its precise path across Italy, the thoughts and reactions of those subjected to its fury, and the precise damage done. Only then did he assess its significance by saying that God wanted to warn rather than punish Tuscany, since had the storm passed over Florence unimaginable disaster would have resulted (6.34). Machiavelli treated the storm much as he had Lorenzo's lightning bolt. Without explicitly saying so, he reminded the reader of the existence of eyewitnesses and gave it veracity by including enough details to make the storm into a visible, datable event. But it is the storm's existence in the consciousness of contemporaries that gives it verisimilitude, making it part of the complex web of actions and fears that gave meaning to fifteenth-century political life and led to the catastrophe of the invasions.

Machiavelli's view that history was grounded in personal realities also played a part in eradicating the distinction between concrete and intangible events. His practice is clear when he had to describe miracles that involved individuals. He found in his sources a story that Stefano Porcari appeared in a dream to Pope Nicholas and announced that he was conspiring to overthrow the papacy. So forewarned in his dream, Nicholas made the necessary preparations and averted the danger. In his own account of this event, Machiavelli changed the nature of the miracle significantly, maintaining only the outer shell of prophecy and modifying all the concrete details to reinforce his interpretation both of Porcari's character and of conspiracies in general.

> He was made hopeful that his plans would succeed by the evil habits of the prelates and the discontent of the barons and the Roman people, but above all he took heart from those verses of Petrarch in the canzone . . . where he says, "On the Capitoline you will see a knight whom all Italy will honor, thinking more of others than of himself." Stefano knew that poets are often endowed with a divine and prophetic spirit, and he thus felt that what Petrarch prophesied in the canzone would happen and that he was the one to carry out so glorious an undertaking, since he was superior in eloquence, learning, charm, and friends to any other Roman. (6.29)

In Machiavelli's hands, divine intervention and prophecy became psychological facts. He saw them as part of Porcari's self-image rather than as an element of Pope Nicholas' actual experience. Machiavelli altered the concrete details of the incident in order to strengthen its verisimilitude, but he still wished to give it a sense of veracity. He buttressed the veracity of the account by quoting a historical document—the *Canzone* of Petrarch—that gave substance to Porcari's conviction that a savior of Italy was destined to come. The story is made to seem both probable and actually true, of general significance and immediate impact. Machiavelli discouraged the reader from separating the narrative into veracious events and probable interpretations. Through the genius of Machiavelli's rhetoric and the profundity of his insight into the implications of Petrarchan time, the narrative became a living process, where truth and probability, veracity and verisimilitude, were inextricably linked.

Intention and will were integral parts of the linear sequence of Petrarchan time, making possible the judgments of historical perspective that were its most splendid fruits. To uncover the linear sequence is to establish an element of meaning. Machiavelli wrote to an audience that accepted this connection between sequence and meaning, and to prove the historicity of his narrative he had to show both that the linear order was meaningful and that it was accurate. These were not sequential steps of a methodical process of verification, where the order comes first and the meaning is derived from it. They were elements of a single gesture in which the reader was convinced of both the meaning and the veracity of the narrative. The solution is thus through rhetoric rather than through logic.

Even in works which were not narrative histories, Machiavelli tended to solve interpretive problems through rhetoric and narrative. His attempt to show in the *Discourses* that virtue rather than fortune was responsible for Rome's rise to power is a good example. Machiavelli based his case on the observation that Rome had deliberately avoided fighting two major powers at once. To prove that the absence of multiple wars was not simply a fortunate circumstance, a modern historian would tend to document Rome's policy by looking for evidence of a conscious intent to fight only one war at a time. Machiavelli accepted Livy's words as an authentic picture of Roman history, and he could have found material to construct such a case in the Roman historian's accounts of senatorial meetings and of other examples of Roman diplomacy. By following this line of inquiry he could establish Roman policy as an actual historical fact capable of being determined with veracity.

Machiavelli did not proceed in this manner, however, for intentions did not lie fully within the field of veracity, and he viewed the problem as one of verisimilitude. To convince his reader of the likelihood of his interpretation, he reconceptualized the temporal sequence, changing it from a lifeless series of events into a living reality with its own dynamics and force. First he gave the view of those who maintained that Rome avoided multiple wars only by luck.[16]

> Those [who emphasize fortune] say that it was due to fortune and not to the virtue of the Roman people that they never fought two major wars at once. Because they did not make war with the Latins until they had beaten the Samnites so badly that the Romans themselves had to make war in their defense. They did not fight with the Etruscans before they had subjected the Latins and completely disheartened the Samnites with frequent routs. Certainly if two of these powers had joined while they were still fresh, it could be conjectured that the destruction of the Roman Republic would have followed.

In this passage Machiavelli portrayed the wars as a series of actions by the Romans. He used transitive verbs to emphasize the activity of the participants and the discrete and isolated quality of the events.

From this summary of his opponents' interpretation of the early wars, he went on to describe them from his own point of view. To portray the sequence of wars as a matter of policy rather than chance, he removed the transitive verbs and substituted ones which portray organic processes of life and death, making of the sequence an extension of personal time and giving to it a substance and coherence that transcends the concrete, mechanical qualities of the first depiction.

> But however it came to be, it never happened that they had two major wars at once. It even appears always that either in the birth of one the other died or that in the death of one the other was born. This can be seen easily in the order of the wars they fought, because . . . while they fought with the Equeans and the Volscians, no other people descended on them. When these were tamed, the war against the Samnites was born. . . . When [the Latins] were subdued, the war with the Samnites revived. When their forces were beaten down . . . the war with the Tuscans was born. (2.1)

In the long passage that follows Machiavelli carried the wars down to the destruction of Macedon and Syria in the East, referring to the wars as "being born" or "reviving."

A superficial reading of this passage might lead to the conclusion that Machiavelli simply repeated in greater detail the sequence he outlined before. But in fact he changed the historical reality from one existing in simple, linear, physical time, to one that existed in the Petrarchan time of personality. The earlier statement of the sequence consisted of separate actions which would have no connection unless an external, visible cause could be identified. In his second description, Machiavelli told the story as an internal one which possessed a dynamic and a coherence by the necessity of its own nature. He had no need to say there was an external causal relationship between the beginning of one war and the end of the previous one; instead he presented the relationship as an organic one in which the parts took on meaning only by the historian's insight into the coherence of the whole.

After this organic description of the sequence of Rome's wars, from those with the Samnites to the wars in the East, Machiavelli explained its internal dynamic to the reader and showed that it consisted in the psychological effect of the series of events. "Anyone who will consider well the order of these wars and the way they proceeded, will see a great virtue and prudence mixed with the fortune. And whoever examines the cause of this fortune will find it easily. Because it is certain that, as a prince and a people acquire so great a reputation that every neighboring prince and people is afraid to attack it, it will always happen that each will only attack if necessary" (2.1). He did not try to prove that the Romans were aware of this strategy and deliberately took advantage of this fear. Instead he simply pointed to the behavior of their enemies, using Carthage as an example, showing that it in fact failed to intervene when Rome was fighting other powers until it was too late.

Machiavelli proved his point here by transforming the issue and recreating the past as an organic thing with a "senso e sapore." By this transformation he obviated the need for specific documentary proof of Rome's intentions. Instead he endowed the entire sequence of Rome's growth with intentionality, dissolving the division between events and words that had dominated classical history. He created thereby a reality whose truth was neither entirely dependent on documentation of visible events existing in a specific time, nor completely a matter of determining the probable motives and strategies behind the visible events. Rome's rise to power is both a series of particular events and a group of strategies and motives. It possesses a reality that comprehends both elements in a coherent unity and that can be expressed only through a rhetorical technique that conveys and accepts that unity.

Three general conclusions stand out from this investigation into the truths of relative time. First, for these historians truth was indeed the daughter of time. The shape of time determined their preconceptions of what is true, how it is discovered, and how it is communicated. The different times which inform the works of major Western historians carry with them important implications for how the writers and their audiences viewed the accuracy and truth of their narratives. Issues of veracity, verisimilitude, faith, and trust are all bound up with the shape of times. Standards of objectivity and truth cannot be understood until we know the times within which the objectivity is sought. This does not mean that the truths are purely subjective, for just as truth is the daughter of time so does the nature of the event determine the shape of time. The truth of the narrative ultimately derives from the reality being described, but this truth takes form only through the specific times that are appropriate to the subject. Truth and time exist in a reciprocal relationship, each defining the other.

Second, the plurality of times that characterized the narratives of historians before Newton created a plurality of standards of historical truth. The twin poles of veracity and verisimilitude dominated the critical methods of these historians, but within each of these categories several different approaches can be found. Historians were comfortable with this variety and used it to enrich and deepen their narratives. They felt that a single standard of accuracy would distort the truth of the past. In particular, they acknowledged a broad range of historical events, including motives and strategies, that were an essential constituent of history but to which they could not apply the same standards they used for concrete events.

Finally, rhetoric had a fundamentally different place in the discovery of truth from the one it occupies in modern life. For most conventional modern historians the relationship between rhetoric and truth is a sequential one. First we discover the truth, the bedrock of locatable facts, through the application of critical method to primary sources. Then with a foundation of objective truth we use rhetoric to shape and embellish our history so that it will be clear, readable, and entertaining. This sequential aspect also obtained among the historians studied here. They accepted at least as much as we do the importance of clear, readable prose and felt that history should be entertaining.

Alongside this sequential relationship, however, was another one that gave to rhetoric a fundamental role in the establishment of truth. To the extent that the realities of the past are not discrete, concrete particles, to the extent that they are organic processes, rhetoric plays a crucial role in discovering and proving the truth, not just in expressing it. Our preconceptions about political and economic reality, about motivation, and about the course and direction of history, all play a role in judgments of truth. Rhetoric gives to these preconceptions a shape and clarity; it discovers the intangible realities that make history a meaningful subject; it makes of the past a living thing. The historians before Newton accepted and understood this role. Though they disagreed on specific issues, they all saw that the historian was deeply involved in his subject and that out of this involvement came whatever usefulness and value history might have.

9

Epilogue

To fashion an identity for a disparate group of events is among the historian's most creative acts. To convince others of that identity and make it a part of their worldview is among the most valuable contributions the historian can make. Our values, our sense of being at home in the world, our self-conception, all grow from a complex network of social, political, and individual identities. The past has played a decisive role in shaping those identities. Most of the historians who have made their mark in Western historical writing have achieved new identities for their contemporaries and expressed those identities with such clarity and force that they formed the vocabulary of the future.

Identity and time have a close and indissoluble relationship. The very simplest act of identification, where we calculate a sum from a group of separate numbers, is made possible only by the linear time in which successive numbers are added to the sum. Kant's insight into the temporal basis of arithmetical judgments, together with the notion of absolute time, led him and his successors to view all identity as a type of synthetic judgment depending on linear time, and hence reducible to a group of separate events on a continuous time line. But historically the process of identity has been more complex, involving a reciprocal relationship with time. Many identities created their own time frames. At the simplest level these time frames were entirely self-contained, like the years of the war Thucydides used to establish the unity of the conflicts between Athens and Sparta. At other levels they transcended the particular identity they were designed to serve and proved useful beyond their immediate scope. The incarnational system served not only to date the Christian commonwealth but also to measure the Petrarchan time of personal and public interaction with the general past.

In this process by which new identities took shape and became part of the subsequent historiographical tradition, more and more types of historical realities were seen as essentially linear in nature. This process created broader and broader instruments of temporal measurement, for linear processes call for a more precise measurement than episodic ones. Herodotus did not see the need to measure the intervals between the events leading to the downfall of the false Smerdis, because he saw two episodic events without specific linear relationship. Thucydides did not measure the intervals between the events leading to the Athenian Empire and Spartan distrust, for these were separate and overlapping episodic processes. He did establish with great precision the temporal relations among the events of the war itself, for he saw the conflict between Athens and Sparta as a single linear series, where antecedent events determined the course of subsequent ones and gave structure and substance to the war.

Within the context of Newtonian time this increase in the scope of linear time seems a development that reflects the nature of time itself, but those who created new linear chronologies did not intend the construction of a universal linear time frame; nor did they see their work as part of this larger process. Their concerns were more particular; they sought only to measure the temporal relationships among events whose identity they had newly perceived and whose linear quality had not been previously apparent. The idea that time itself might be linear was not present. Yet the process itself of developing new linear chronologies had a linear dimension to it. Once some aspect of reality was measured in linear time, subsequent historians were not again likely to see that aspect in episodic terms or to abandon the capacity for precise temporal location that linear measurement afforded.

There also seems to be an order governing the type of processes that become linear. At the beginning two qualities gave certain historical processes linear dimensions: physical identity and consciousness. Herodotus saw accretion of the land of Egypt as a linear process because it had a clear physical identity, one that was visible to him in the vivid contrast that still exists in Egypt between the Nile valley and delta on the one hand and the desert on the other. Changes in the shape of this land seemed obviously cumulative to him and demanded a measurement of time that differed from that he used for political and military processes. By the same token the learning of techniques of religious worship or shipbuilding had an identity it drew from the transmission of conscious knowledge from one person or country to the next. Here too the cumulative nature of this knowledge seemed obvious and suggested the advisability of a linear measurement.

Political identity was more problematic. Though there were inescapable elements of consciousness and geography in the exercise of political power, there were also qualities of intentionality and spirituality that for the Greek historians created a process too complex to be entirely linear. Thucydides brought one aspect of political identity into linear focus, when he saw that armed conflict could be presented as a sequence of events with a beginning and end. Here again warfare has not only a physical nature but an element of conscious strategy and technique that makes it easier to incorporate into a linear framework than the more subtle processes by which political domination acquired legitimacy. There Thucydides tended to emphasize the irrational and the private. He avoided a picture of political change where political consciousness made antecedent strategies an ineluctable element in consequent results. Describing Athens' liberation from tyranny, he stressed the private motives of the conspirators to avenge the insults arising from a love affair, deliberately playing down the issue of political freedom (6.53–61).[1]

The meteoric political and military intervention of Rome into political life of the Mediterranean gave to that process the qualities of linear time. Even there, the new linear chronology defined that specific conquest and unification, leaving out other political processes that had no direct relation to it. Many centuries passed before the history of political legitimacy itself acquired a linear quality and an appropriate chronology. Pompeius Trogus made linear the process by which political legitimacy moved from one state to another, but he did not measure it and left out many important political changes. It took the Christian Orosius to provide an appropriate measure—from the founding of Rome—around which all important secular political changes could be organized into a single linear series.

Aspects of experience where will and intention were more intrinsic remained recalcitrant to linear measurement. The growth of Christianity and the history of the Christian commonwealth took on linear dimensions and acquired a distinctive chronology, dating from the birth of Christ. In using this chronology, however, historians eliminated many real issues of will and spirituality and tended to treat the rise of Christian states as a purely political process, parallel to the growth of secular states in the pre-Christian period. The idea of the two cities remained a theoretical construct, without real application in the practice of historical narrative.

Personal development remained outside this gradual accretion of linear chronologies. Though individuals possess a clearly defined physical identity and are in fact the locus of consciousness, the element of will so crucial to personal identity defied incorporation into a linear time line. Thus classical and medieval historians portrayed personal development in complex and contradictory ways that resisted inclusion in the linear time frames they developed for other historical realities. Only the personalization of all historical processes by the great Renaissance historians eliminated the complexity of different time frames from their narratives. With a new sense of the identity of the past as a whole, Machiavelli and Guicciardini treated all of history as though an analysis of linear relationships could provide insight into the meaning of events.

Personal identity is the most obvious area where absolute time impedes the implementation of a historical method for validating one's perceptions, but the problem of validation remains in other areas of identity. In the twentieth century it has become increasingly clear that cultural, social, and even political identities suffer distortion as they are reduced to simple elements and placed in a single linear series. I will conclude this study with some reflections on three representatives of modern thought whose studies of identity have led them into paths which challenge standard historical methodologies. They have approached this problem from perspectives which accept the discontinuities of modern time and have sought a basis for identity through historical methods that differ significantly from those commonly accepted among modern historians. Instead they have adopted methods more consonant with those used by the historians studied here. Two of them, Marcel Proust and Michel Foucault, are quite self-conscious in their rejection of Newtonian time and space; they are sensitive to the existence of new times in this century and aware that means of verification are necessary to establish the identity of events within these new times. The third, Erik Erikson, is less self-conscious about the new times, but in his attempt to discuss personal identity from a historical point of view, has nonetheless adopted a methodology that resembles that of the other two.

We are so accustomed to apply absolute time to the quotidian details of life that we do not readily appreciate the extent to which it changes the focus of identity, depersonalizing historical realities and removing them from immediate contact with life. The scientific framework of which absolute time is a part ascribes identity on the one hand to the absolute thinking subject, removed from all historical development, and on the other to the unchanging primary qualities of material things, including time and space. As a means of measuring the intervals between concrete events in terms of these primary qualities, the absolute dating system is implicitly subversive of traditional notions of personal identity. Hume used it effectively to challenge the very existence of personality.

> As a memory alone acquaints us with the continuance and extent of this succession of perceptions, it is to be considered upon that account chiefly, as the source of personal identity. . . . But having once acquired this notion of causation from memory, we can extend the same chain of causes, and consequently the identity of our persons beyond our memory, and can comprehend times, and circumstances, and actions, which we have entirely forgot, but suppose in general to have existed. For how few of our past actions are there of which we have any memory? Who can tell me, for instance, what were his thoughts and actions on the first of January 1715, the 11th of March 1719, and the 3rd of August 1733? Or will he affirm, because he has entirely forgot the incident of these days, that the present self is not the same person with the self of that time; and by that means overturn all the most established notions of personal identity? In this view, therefore, memory does not so much produce as discover personal identity, by showing us the relation of cause and effect among our different perceptions. It will be incumbent on those who affirm that memory produces entirely our personal identity, to give a reason why we can thus extend our identity beyond our memory. (*Treatise*, 1.4.6)

Hume's question would be entirely without force if posed in one of the dating systems discussed in this book. It would be meaningless to ask Augustine what he was thinking about on the 15th of the Kalends of March in the 1,155 year from the founding of the City, for that date applied to a set of political phenomena which were not universal and which did not necessarily include the process of his own personal development. Hume could not even challenge Augustine with gaps in his memory, for the gaps occurred only in consciousness. Memory provides the continuity by definition; it does not depend on an external framework of continuous time.

The question has also lost its force in the twentieth century, where time is no longer an external, absolute continuum but a discontinuous process enmeshed in the act of perception and measurement. Within such a context identity does not depend on the continuity of memories in an external framework; it comprehends not only the discontinuity of particular memories but that of time itself. To accommodate the discontinuities of modern time new conceptions of identity and new standards of truth have arisen, meeting the challenges new scientific concepts have posed to the unity of personality. Literature and art have led the way, partly because there the issue of identity takes on a specificity it often lacks in scientific and psychological theory. Novelists, poets, and artists face directly the problem of making the individuals in their work seem true and probable. As conceptions of time have changed in the twentieth century, writers often have had to explain their perspective to a readership still tied to Newtonian science. Thus literature in this century can serve as a methodological source to illumine the problems of making believable a presentation of individual identity in contemporary time, offering to historians a model for the realization of present day experience.

Many twentieth century writers, from Joyce and Mann to Borges, Vargas Llosa, and Pynchon, have developed the ramifications of the new time, but none has probed the relation of time and memory more deeply than Marcel Proust, whose protagonist, in *Remembrance of Things Past*, explored his own memories and sought to recover the past that he had lost in the process of living his life. In one of the most well-known passages of modern literature, the adult narrator recovers a scene from his childhood when the taste of a madeleine dipped in tea reminds him of a similar cookie given him by his Aunt Léonie at Combray, where the family house in the country was located. This memory sends him on a long search for his past, which culminates in the scene at the end where the loose paving stones in a Parisian courtyard bring back a similar sensation from Venice in years past, which in turn causes his whole past to flood back into his consciousness and shows him the world as it has actually changed in the time during which he has lived.

To portray this search Proust dissolved the conventional unities of subject and object. The narrator's very identity seems to lose itself, becoming more fractured and dissociated the more closely he focuses on it. "Our ego is composed of the superimposition of our successive states. But this superimposition is not unalterable like the stratification of a mountain. Incessant upheavals raise to the surface ancient deposits."[2] These changes in the ego have neither linear order in time nor logical structure.

> I had always considered each one of us to be a sort of multiple organism or polyp, not only at a given moment of time—so that when a speck of dust passes it the eye, an associated but independent organ, blinks without having received an order from the mind, and the intestine, like an embedded parasite, can fall victim to an infection without the mind knowing anything about

it—but also, similarly, where the personality is concerned and its duration through life, I had thought of this as a sequence of jux-taposed but distinct "I's" which would die one after the other or even come to life alternately. (3.972)

This perspective would seem to destroy the very identity that Proust was so determined to recover. The individual dissolves into component parts which have relation only to the larger pattern of nature in general. But when he treated this overall pattern of nature, Proust made it in turn de-pend on the identity of the individual observer. "Poets claim that we re-capture for a moment the self that we were long ago when we enter some house or garden in which we used to live in our youth. But these are most hazardous pilgrimages, which end as often in disappointment as in success. It is in ourselves that we should rather seek to find those fixed places, con-temporaneous with different years" (2:89). Because of the power of the imagination over reality, each of us creates a universe around us that can-not be effectively tested against some "general" universal reality.

The universe is real for us all and dissimilar to each one of us. If we were not obliged, in the interests of narrative tidiness, to con-fine ourselves to frivolous reasons, how many more serious rea-sons would enable us to demonstrate the mendacious flimsiness of the opening pages of this volume in which, from my bed, I hear the world awake, now to one sort of weather, now to another! Yes, I have been forced to whittle down the facts, and to be a liar, but it is not one universe, but millions, almost as many as the number of human eyes and brains in existence, that awake every morning. (3:189–90)

Reality for Proust lies in this complex connection between a personal memory—as arbitrary in its power as the imagination (2:384)—and an external world, which includes other peoples' memories. What Newton or Hume might have called reality—the visible events and abstract con-structs that allow communication, analysis, and social life—Proust thought no more than "the waste product of experience" (3:925). Absolute mea-sures of time and space were insufficiently real and missed the most impor-tant elements of experience. "An hour is not merely an hour, it is a vase full of scents and sounds and projects and climates, and what we call real-ity is a certain connexion between these immediate sensations and the memories which envelop us simultaneously with them—a connexion that is suppressed in a simple cinematographic vision, which just because it

professes to confine itself to the truth in fact departs widely from it"
(3:924).

Proust's warnings about the arbitrariness of memory and the illusionary
qualities in snapshot visions of past time are not meant to diminish the
role of time in life or to free the individual from the influences of the past.
Quite the contrary, few writers have felt so deeply the force of time. For
Proust, time created our personality and delimited our responses to new
stimuli. It is precisely because memory often suppresses the role of time
that its pictures are inaccurate. "Memory by itself, when it introduces the
past, unmodified, into the present—the past just as it was at the moment
when it was itself the present—suppresses the mighty dimension of Time
which is the dimension in which life is lived" (3:1087).

Time forms our personalities not only as we experience life directly but
also as we take on the personalities of those who were early influences on
us, especially parents.

> Little by little, I was beginning to resemble all my relations. . . .
> Although every day I found an excuse in some particular indis-
> position, what made me so often remain in bed was a person . . .
> who had transmigrated into me . . . and that person was my aunt
> Léonie. . . . When we have passed a certain age, the soul of the
> child that we were and the souls of the dead from whom we
> sprang come and shower upon us their riches and their spells, ask-
> ing to be allowed to contribute to the new emotions which we
> feel and in which, erasing their former image, we recast them in
> an original creation. Thus my whole past from my earliest years,
> and beyond these, the past of my parents and relations, blended
> with my impure love for Albertine the tender charm of an af-
> fection at once filial and maternal. (3:72)

The insight that his parents' past was part of his own personality under-
cuts Hume's objection, for Proust saw in the gaps of his memory no threat
to his identity. He boldly proclaimed that his identity persevered beyond
the limits not only of his consciousness and memory but of his physical
life.

> What, then, is a memory which we do not recall? Or, indeed, let
> us go further. We do not recall our memories of the last thirty
> years; but we are wholly steeped in them; why then stop short at
> thirty years, why not extend this previous life back to before our
> birth? If I do not know a whole section of the memories that are
> behind me, if they are invisible to me, if I do not have the faculty

of calling them to me, how do I know whether in that mass that is unknown to me there may not be some that extend back much further than my human existence? If I can have in me and round me so many memories which I do not remember, this oblivion (a *de facto* oblivion, at least, since I have not the faculty of seeing anything) may extend over a life which I have lived in the body of another man. (2:1017–18)

Since the past was so crucial to determining identity, Proust saw the importance of presenting it accurately. He sought a true picture of the past and knew that his findings must somehow be tested against reality. He also understood that the will was the source of bias, with its inevitable tendency to distort reality in the observer's interest.

It is one of the faculties of jealousy to reveal to us the extent to which the reality of external facts and the sentiments of the heart are an unknown element which lends itself to endless suppositions. We imagine that we know exactly what things are and what people think, for the simple reason that we do not care about them. But as soon as we have a desire to know, as the jealous man has, then it becomes a dizzy kaleidoscope in which we can no longer distinguish anything. (3:529)

He condemned as ineffective the traditional means of eliminating that kind of bias, for they assumed an absolute time line and a subject that could be separated from the events. Proust found truth not in separating himself from the events but in embracing them, accepting them as they occurred. He removed his will not so much from the act of perception—which was impossible—as from the act of analysis and criticism. At the end of his novel he discussed the process by which he authenticated the sensations and associations which had brought back his past and confirmed his identity. He asserted that the mark of truth lay above all in finding associations which we have not sought. "For the truths which the intellect apprehends directly in the world of full and unimpeded light have something less profound, less necessary than those which life communicates to us against our will in an impression which is material because it enters us through the senses but yet has a spiritual meaning which it is possible for us to extract" (3:912).

Proust's method avoids deliberate testing of hypotheses by experience, but it does not resort to passive receptivity. Though he does not actively connect the sensation with an unchanging reality, his activity is one of drawing from the sensation the exact spiritual quality which it evoked in his own consciousness and the memories associated with it. With Proust, as with Vico, art became the source and criterion of truth.

> And this method, which seemed to me the sole method, what was it but the creation of a work of art? . . . Their [the memories'] essential character was that I was not free to choose them, that such as they were they were given to me. And I realised that this must be the mark of their authenticity. I had not gone in search of the two uneven paving stones of the courtyard upon which I had stumbled. But it was precisely the fortuitous and inevitable fashion in which this and other sensations had been encountered that proved the trueness of the past which they brought back to life, of the images which they released, since we feel, with these sensations, the effort that they make to climb back towards the light, feel in ourselves the joy of rediscovering what is real. (3:912−13)

Proust's method embraced the complexities and discontinuities of modern time. In particular it bridged the gap between unconscious and conscious realities that Newtonian time tends to make so rigid and absolute. The identification of truth became not simply a conscious analysis of static events but an inquiry into the unconscious bases of experience and perception. He thought the conscious and unconscious selves existed in different times, and though his method established their connection, they remained separate and needed different treatment. Pleasures and feelings have the same sense in the two different states. Like Freud, Proust turned to the orgasm to explore the different times of the self and the complexity of our perception of pleasure.

> One can of course maintain that there is but one time, for the futile reason that it is by looking at the clock that one established as being merely a quarter of an hour what one had supposed a day. But at the moment of establishing this, one is precisely a man awake, immersed in the time of waking men, having deserted the other time. Perhaps indeed more than another time: another life. We do not include the pleasures we enjoy in sleep in the inventory of the pleasures we have experienced in the course of our existence. To take only the most grossly sensual of them all, which of us, on waking, has not felt a certain irritation at having experienced in his sleep a pleasure which, if he is anxious not to tire himself, he is not, once he is awake, at liberty to repeat indefinitely during that day. It seems a positive waste. We have had pleasure in another life which is not ours. (2:1015)

Proust's method for verifying his past, since it did not depend on connecting sensations and feelings through an external, linear time line, produced interrelationships among events which transcended conventional temporal order. Though they violated Newtonian conceptions of the temporal order of cause and effect, Proust felt they conveyed realities whose force was undeniable. At one point the narrator opens by mistake a letter he thinks is addressed to his mistress and finds an assignation written in code. He immediately suspects that the letter was written by his own mistress and was evidence of her infidelity. He soon discovers that he is mistaken and that the letter was actually written to another person entirely, with whom he has no connection.

> And so I had on that occasion been utterly mistaken in my suspicions. But the intellectual structure which had linked these facts, all of them false, together in my mind was itself so strict and accurate a model of the truth that when, three months later, my mistress (who had at that time been meaning to spend the rest of her life with me) left me, it was in a fashion absolutely identical with that which I had imagined on the former occasion. A letter arrived, containing the same peculiarities which I had wrongly attributed to the former letter, but this time it was indeed meant as a signal. (3:433)

In this case the first letter had created a true picture of the narrator's future relations with his mistress, though it flatly contradicted her conscious intentions at the time it was written. Truth depends not on the relationship between individual concrete events and an omniscient, separated observer, but on the context and perspective of the participant. Later on in the work, when Proust is considering the process by which he became a writer and the relation between early intentions and later results, he observes, "In this context certain comparisons which are false if we start from them as premises may well be true if we arrive at them as conclusions" (3:936).

Proust's method of verification has many affinities with the techniques of verisimilitude which formed part of the historians' craft for many centuries before Ranke. This method did not focus exclusively on the relation between the particulars of the narrative and some unchanging, external reality, where truth values were objective and existed independently from the observer and the reader. Instead it stressed the relation of sensations and feelings within the observer, structuring these in such a way that the reader would find them convincing. For Proust, as for Thucydides, Tacitus, and Machiavelli, the historian's craft was in the last analysis an activity of art more than of science. It involved the imaginative recreation of reality more than the impartial sifting of facts that were true from any perspective. Proust and the historians both found a method of verification which allowed them to present a picture of personal identity which transcended and included the discontinuities of time. In these narratives, personal realities were not reduced to primary facts existing on a continuous time line.

Proust may seem a bizarre model to hold up to historians. Yet identity, and its connection with the social experience that occupies so much of his attention, are realities that few historical methods based on absolute time or scientific analysis can successfully accommodate. Here even the most rigid of traditional historians must use techniques that seek verisimilitude rather than veracity. Since verisimilitude by its very nature depends at least in part on the preconceptions of the readers, Proust is not so easily ignored by historians. As the narrative techniques and temporal dimensions in which he pioneered become more and more the stock-in-trade of working novelists and writers, traditional pictures of personal development will seem less and less "real" to readers. To present individuals in ways that seem convincing to their readers, historians will increasingly have to shape personality in terms more like those of Proust and Calvino—and, coincidentally, of Suetonius and Bede—than those of Dickens and Eliot.

The relation of personal identity to social and political forces has not been an easy one for historians to resolve, though its importance is such that no one can claim to study the past with any depth who does not have some model for this relationship. In describing the activity of individuals, there must be some set of preconceptions, whether self-conscious and explicitly argued or implicit and hidden among the unconscious assumptions that all bring to their work. Furthermore, attempts to substantiate the personal development of historical figures have often produced techniques of verification which hearken back to the verisimilitude of the pre-modern historical tradition.

A particularly well-known treatment of identity is Erik Erikson's *Young Man Luther*, which, appearing in 1958, introduced many of the concepts and techniques of that branch of history known as psychohistory. Erikson is not himself a professional historian but a psychoanalyst, and he approached his topic not simply with a view to describing the past as it was but in order to derive use and value from it. "To relegate Luther to a shadowy greatness at the turbulent conclusion of the Age of Faith does not help us see what his life really stands for. . . . Historical analysis should help us to study further our own immediate tasks, instead of hiding them in a leader's greatness."³ Erikson saw Luther as one who opened up new paths of introspection for the sixteenth century much as Freud did for our own. He focused on three aspects of Luther's thought: "the affirmation of voice and word as the instruments of faith; the new recognition of God's 'face' in the passion of Christ; and the redefinition of a just life" (p. 207). Erikson argued that these constituted a major redefinition of the human condition, a redefinition which "while part and parcel of his theology— has striking configurational parallels with inner dynamic shifts like those which clinicians recognize in the recovery of individuals from psychic distress. In brief, I will try to indicate that Luther, in laying the foundation for a 'religiosity for the adult man,' displayed the attributes of his own hard-won adulthood; his renaissance of faith portrays a vigorous recovery of his own ego-initiative" (p. 206).

Erikson's search for usefulness and contemporary relevance in the past led him to pose an intimate connection between Luther's personal identity and the major ideas of the Protestant Reformation. The history of Luther's personal psychological development thus took on a general and public significance. If the process by which Luther found his identity played a role in the theology through which he expressed it, then Erikson confronted a problem that was unavoidably historical, and that cannot be solved solely by theoretical constructs. He needed to study those particulars of the past that would illumine the unconscious processes of the self as they give shape to desire. Erikson accepted this challenge gladly, feeling that traditional historical approaches failed to uncover the truth here in a meaningful and useful fashion.

> We cannot leave history entirely to nonclinical observers and to professional historians who often all too nobly immerse themselves into the very disguises, rationalizations, and idealizations of the historical process from which it should be their business to separate themselves. Only when the relation of historical forces to the basic functions and stages of the mind has been jointly charted and understood can we begin a psychoanalytic critique of society as such without falling back into mystical or moralistic philosophizing. (pp. 20–21)

The exploration of Luther's past that constitutes the core of Erikson's book is complex and subtle, but one example will serve to illustrate the distinctive features of his approach. He was convinced that behind Luther's adult theology lay an arrested early development, in which deep forces and influences prevented Luther from accepting the identity which his society had prepared for him and which so many of his colleagues had so easily assumed. As part of his evidence for this arrest, Erikson adduced the so-called "fit in the choir."

> Three of young Luther's contemporaries (none of them a later follower of his) report that sometime during his early or middle twenties, he suddenly fell to the ground in the choir of the monastery at Erfurt, "raved" like one possessed, and roared with the voice of a bull: "*Ich bin's nit! Ich bin's nit!*" or "*Non sum! Non sum!*" The German version is best translated with "It isn't me!" the Latin one with "I am *not!*" (p. 23)

The fit in the choir is not a primary fact, less because the sources are so late and so vague than because it has no existence apart from Luther's psychic development. It cannot be located simply in time, like the day he entered the monastery or the day he stood before the emperor and said his conscience would not permit him to go against the word of God. Such events as those exist in public time; they show Luther in conflict or conjunction with the social institutions of his day. They have a clear physicality and conscious institutional symbolism that offer the possibility of a precise and absolute date. The fit occurred only in context with the process of growth and the struggle for identity that culminated in the mature expression of his theology and life. Its physicality, the fact that it was a visible event within a social context, allowed it to become part of the historical record, but even then it took on historical status only as a perception among his colleagues, not as an event in its own right.

Erikson accepted these qualities of the event and insisted that other standards of verification be applied to it.

> Judging from an undisputed series of extreme mental states which attacked Luther throughout his life, leading to weeping, sweating, and fainting, the fit in the choir could well have happened; and it could have happened in the specific form reported, under the specific conditions of Martin's monastery years. If some of it is legend, so be it; the making of legend is as much part of the scholarly rewriting of history as it is part of the original facts used in the work of scholars. We are thus obliged to accept half-legend as half-history, provided only that a reported episode does not contradict other well-established facts; persists in having a ring of truth; and yields a meaning consistent with psychological theory. (p. 37)

In this passage Erikson has proposed a series of techniques aimed at producing verisimilitude rather than veracity. Moreover, these techniques are consonant with those used by the historical tradition from Herodotus to Guicciardini to establish the truthfulness of those parts of their history for which veracity was inappropriate. Erikson was not simply establishing the probability of the event; he thought it actually happened. But he did realize that arguments from probability, together with a sound theoretical apparatus, must be used in order to create a "ring of truth" for a story which has no existence as a primary fact.

To approach such a problem he had to concentrate less on the relation of the event to the data than on the ways historians have tried to explain the data to their readers. To show that the evidence of the fit in the choir has stimulated diverse reactions among those who have read it he analyzed three Luther scholars and their treatment of the fit. Erikson found it significant that, despite their skepticism, each was extraordinarily fascinated with the story and—whether consciously or not—each used it in important ways to shape his picture of Luther.

> For with the same facts (here and there altered, as I have indicated, in details precisely relevant to psychological interpretation), the professor, the priest, the psychiatrist, and others as yet to be quoted each concocts his own Luther. This may well be the reason why they all agree on one point, namely, that dynamic psychology must be kept away from the data of Luther's life. Is it possible that they all agree so that each may take total and unashamed possession of him, of the great man's charisma? (pp. 29–30)

Erikson has turned here the very skepticism and rejection among scholars into a sign of the story's inner truth, of its ability to conjure up a strong sense of significance and awaken the deep recesses where Luther's identity touches our own.

Erikson approached Luther's biography as an outsider to the profession, stimulated by the desire to find significance and apply his findings to life. His work has attracted much skepticism, but in the nearly twenty years since its appearance the proposition that some understanding of the psychic processes in personal development is important has become generally accepted. What is less accepted is the need Erikson acknowledged to modify traditional historical method in pursuit of data which were not amenable to its treatment, data which did not testify to a primary fact locatable in time and space without reference to other facts.

Personal identity, however, is only the most obvious among the aspects of historical reality that resist incorporation into absolute time. Cultural and social identity are also problematic. Changes in these areas can seldom be reduced to groups of primary facts, though the visibility of writing often gives to the works of philosophy and literature that symbolize such identity a superficial simplicity of location. Another outsider to the historical profession, Michel Foucault, devoted much of his career to exploring this problem. Trained in philosophy and psychology, he wrote the history of madness, of prisons, and most recently, of sexuality. In his hands these phenomena become symbols of deeper relations, especially of the deployment of power and the alterations in modes of discourse that permit the exercise of power. He portrays in these relations of power a complex and discontinuous interaction of institutions and individuals, of language and action, of power and thought.

Among Foucault's most impressive achievements is a study of the changes that occurred in certain areas of scientific thought in the period from the sixteenth to the twentieth centuries. In that work Foucault demonstrated an extraordinarily acute insight into the various senses of time that different periods of intellectual history have displayed. He realized that the seventeenth and eighteenth centuries, which he called the Classical Age, were a period firmly based on concepts of absolute and continuous time. He argued that a new type of time underlay the new sciences that arose in the modern period.[4]

> We perceive that Classical thought related the possibility of spatializing things in a table to that property possessed by pure representative succession to recall itself on the basis of itself, to fold back upon itself, and to constitute a simultaneity on the basis of a continuous time: time became the foundation of space. In modern thought, what is revealed at the foundation of the history of things and of the historicity proper to man is the distance creating a vacuum within the Same, it is the hiatus that disperses and regroups it at the two ends of itself. It is this profound spatiality that makes it possible for modern thought still to conceive of time—to know it as succession, to promise it to itself as fulfillment, origin, or return.

Since conceptions of time itself are involved in the changes Foucault wanted to study, he approached his subject with a lively suspicion of continuity. "Confronted by such a curious combination of phenomena, it occurred to me that these changes should be examined more closely, without being reduced, in the name of continuity, in either abruptness or scope. . . . It seemed to me, therefore, that all these changes should not be treated at the same level, or be made to culminate in a single point" (p. xii). Foucault was not simply respecting the complexity of the evidence; he was also determined not to distort the history of thought by enforcing on it a linear order. Foucault implemented this approach throughout his work, insisting on a fresh perspective and avoiding simple relations of cause and effect based on a linear order. After identifying the common elements of the Classical mode of knowledge, he observed,

> At the level of the history of opinions, all this would appear, no doubt, as a network of influences in which the individual parts played by Hobbes, Berkeley, Leibniz, Condillac, and the "Idéologues" would be revealed. But if we question Classical thought at the level of what, archaeologically, made it possible, we perceive that the dissociation of the sign and resemblance in the early seventeenth century caused these new forms—probability, analysis, combination, and universal language system—to emerge, not as successive themes engendering one another or driving one another out, but as a single network of necessities. And it was this network that made possible the individuals we term Hobbes, Berkeley, Hume, or Condillac. (p. 63)

He thus treated his story episodically and in fact used conventional dates less as points on a time line than as boundaries of episodes within which the precise chronological relationship had no real significance.

> The constitution of so many positive sciences, the appearance of literature, the folding back of philosophy upon its own development, the emergence of history as both knowledge and the mode of being of empiricity, are only so many signs of a deeper rupture. Signs scattered through the space of knowledge, since they allow themselves to be perceived in the formation here of philology, there of economics, there again of biology. They are chronologically scattered too: true, the phenomenon as a whole can be situated between easily assignable dates (the outer limits are the years 1775 and 1825); but in each of the domains studied we can perceive two successive phases, which are articulated one upon the other more or less around the years 1795–1800. (p. 211)

Changes occurred at different rates and required different chronologies in the various disciplines Foucault studied. Speaking of developments in the field of language, he said, "They take a more discreet form and obey a slower chronology than in the field of natural history" (p. 232). To find an identity among changes that were apparently unrelated but simultaneous within a period of years, Foucault realized, he could not assume the intrinsic linear order that the traditional historical approach had used to interpret change.

Rejecting the traditional concepts of time and space, Foucault also dismissed the notion of the absolute, thinking subject as the precondition of truth.

> If there is one approach that I do reject, however, it is that . . . which gives absolute priority to the observing subject, which attributes a constituent role to an act, which places its own point of view at the origin of all historicity—which, in short, leads to a transcendental consciousness. It seems to me that the historical analysis of scientific discourse should, in the last resort, be subject, not to a theory of the knowing subject, but rather to a theory of discursive practice. (p. xiv)

Instead of transcendental consciousness Foucault sought the unconscious levels where the preconditions of thought take shape. Historians of science often look on the unconscious as a negative area that impedes the exploration of scientific developments and increases the difficulty of explaining them, but Foucault argued for a "positive unconscious," which would move the historian closer to the scientific discourse by identifying commonalities that were not part of the conscious world of scientific inquiry (p. xi).

Foucault looked upon his endeavor as an archaeology, one which sought the roots from which changes in scientific method and disciplines become possible. As such he was very much concerned with the temporal aspect of his discipline, but his time was that of Vico and Proust rather than of Newton and Kant. He shared Vico's awareness that the origins themselves shape the time in which they are perceived, and he agreed with Proust that the process of finding origins is a reciprocal one. "It is always against a background of the already begun that man is able to reflect on what may serve for him as an origin. For man, then, origin is by no means the beginning—a sort of dawn of history from which his ulterior acquisitions would have accumulated. Origin, for man, is much more the way in which man in general, any man, articulates himself upon the already-begun of labour, life, and language" (p. 330).

Origin played for Foucault the role that sleep played for Proust in creating that connection between our conscious selves and the other times in which we all live. "Paradoxically, the original, in man, does not herald the time of his birth, or the most ancient kernel of his experience; it links him to that which does not have the same time as himself; and it sets free in him everything that is not contemporaneous with him" (p. 331). Thus the search for origins is a profoundly radical inquiry into the bases of consciousness, challenging our deepest assumptions, orienting us to a new and more profound view of time. As a result of such an exploration we should see time not as an endless and continuous series but as a process with its own origins and archaeology. "Such a task implies the calling into question of everything that pertains to time, everything that has formed within it, everything that resides within its mobile element, in such a way as to make visible that rent, devoid of chronology and history, from which time issued" (p. 332).

Foucault saw that the origins of modern economic, biological, and linguistic theory could not be reduced to a series of linear events, just as Thucydides presented the origins of the Peloponnesian War in terms that were not strictly linear. In both cases the superficial, visible events could be placed in linear time. Thucydides certainly knew the date of the Athenian expedition to Thasos, though we cannot tell if he knew the precise date of such other events as the siege of Ithome. With Foucault the case is much clearer. All of the works he discussed can be precisely dated, and in most cases the dates on which their authors came in contact with specific writers or experiences that shaped their opinions can be identified. But such a perspective in Foucault's mind would obscure the principal change he wished to study. Changes in sensibility underlay all the fields he took under consideration, occurring simultaneously within each. Though the changes were simultaneous, there was no conscious influence of one upon the other. The identity he pursued defied linear order and existed in another time, just as the identity Thucydides sought lay in the simultaneous Athenian growth and Spartan fear, not in the linear order of visible events that made them manifest.

In pointing to dimensions of scientific change which absolute time does not accommodate, Foucault challenged traditional historiography at one of its most secure points. Ever since Herodotus such change had been seen as linear, and it had as easily fit into the absolute time as did political and military history. But even activities so dominated by consciousness as scientific inquiry contain elements of unconscious will and preconceptions over which the scientist has no direct control. Such aspects of the discipline do not change in simple linear ways, and Foucault offered an alternative means of investigation that he claimed would bring out the true features of change in science.

From the individual to the scientific community, identity takes a form that cannot be entirely comprehended within the context of a universal, linear time frame. Some have even challenged the long claim of political history. Jacques Le Goff, of the Annales School has suggested that the longue durée also reaches into political history, where changes have elements of deep, not wholly conscious, archetypes that are obscured by treating the surface events of institutional and military change.[5] The work of these scholars is gradually opening up broader and broader areas of human experience to the temporal and spatial framework that has come into being in the twentieth century. The shape that human experience will take when it is fully incorporated into the new time is not yet clear, any more than the implications of the scientific revolution were clear in the sixteenth century, when it had barely begun. What is clear is that the new space and time can no longer be confined to those fringes of experience—the vast distances of outer space or the limited world of subatomic particles—that so many of us would relegate them to. They will have an impact on life itself. And, given the affinities with the past that have been suggested here, the new conceptions of space and time also offer the possibility of reawakening that connection with the premodern world that has been been so weakened by contemporary changes in science and technology.

The stakes in such an endeavor are high, for they speak to the relevance of history to life and the force with which the past may guide its students in meeting the challenges of the future. At an even deeper level, they change the very definition of life by revealing the dehumanizing and deadening qualities of Newtonian time. At the end of Calvino's novel Mr. Palomar finds himself concerned with death, and wonders if he can use the indivisibility of linear time to hold off death by dividing his moments into an endless series.

> "If time has to end, it can be described, instant by instant," Mr. Palomar thinks, "and each instant, when described, expands so that its end can no longer be seen." He decides he will set himself to describing every instant of his life, and until he has described them all he will no longer think of being dead. At that moment he dies.

Notes

1. Italo Calvino, *Mr. Palomar*, trans. William Weaver (New York: Harcourt Brace Jovanovich, 1985), p. 3.

2. Stephen Kern, *The Culture of Time and Space* (Cambridge, Mass.: Harvard University Press, 1983), has presented a clear and cogent picture of this change.

3. See Chapter 2 for a discussion of the role of absolute space and time in his work.

4. Marc Bloch, *The Historian's Craft*, trans. Peter Putman (Manchester: Manchester University Press, 1954), pp. 27–28.

5. In *Varieties of History*, ed. Fritz Stern, (Cleveland: World, 1956), p. 249.

6. In *The Philosophy of History in Our Time*, ed. Hans Meyerhoff (Garden City, N. Y.: Doubleday, 1959), p. 87.

7. E. J. Bickerman, *Chronology of the Ancient World*, 2d ed. (London: Thames and Hudson, 1980), p. 9.

8. Johannes Burkhardt, *Die Entstehung der modernen Jahrhundertrechnung* (Groppingen: A. Kümmerle, 1971).

9. See for instance Stephen Toulmin and June Goodfield, *The Discovery of Time* (New York: Harper and Row, 1965), pp. 20–21.

10. Fernand Braudel, *The Mediterranean and the Mediterranean World in the Age of Philip II*, trans. Siân Reynolds, 2 vols. (New York: Harper & Row, 1972), 1:20–21.

11. Braudel, *On History*, trans. Sarah Matthews (Chicago: University of Chicago Press, 1980), p. 29.

12. Braudel, *The Structures of Everyday Life: The Limits of the Possible*, vol. 1 of *Civilization and Capitalism: 15th–18th Century*, trans. and rev. Siân Reynolds, (New York: Harper & Row, 1981), p. 25. For a philosopher's perspective on Braudel and the *Annales* see Paul Ricoeur, *Time and Narrative*, trans. Kathleen

McLaughlin and David Pellauer, 3 vols. (Chicago: University of Chicago Press, 1983–), 1:99–111, 209–17.

13. Benjamin Whorf, *Language, Thought, and Reality*, ed. J. Carroll (New York: Technology Press of the Massachusetts Institute of Technology, 1956).

CHAPTER TWO

1. Cited in *The Varieties of History*, ed. Fritz Stern, p. 57.

2. René Descartes, *Discourse on Method*, trans. D. Cress (Indianapolis and Cambridge: Hackett, 1980), p. 10. *Oeuvres de Descartes*, ed. C. Adam and P. Tanner, 12 vols. (Paris: Cerf, 1897–1913), 6:19.

3. Descartes, *Meditations*, trans. L. Lafleur (New York: The Liberal Arts Press, 1951), pp. 43–44; *Oeuvres*, 9:39.

4. Descartes, *Les Principes de la philosophie*, pt. 1, sec. 57; *Oeuvres*, 9:49.

5. Isaac Newton, *Mathematical Principles of Natural Philosophy*, trans. A. Motte, rev. F. Cajori (Berkeley: University of California Press, 1960), bk. 1, def. 8, scholium; p. 6. Newton drew his concept partly from the English Platonist Henry More.

6. Frank Manuel, *Isaac Newton Historian* (Cambridge, Mass.: Harvard University Press, 1963), p. 10.

7. John Locke, *An Essay Concerning Human Understanding* (Oxford: The Clarendon Press, 1975), bk. 2, chap. 13, sec. 2–4.

8. David Hume, *A Treatise on Human Nature* (Baltimore: Penguin, 1969), bk. 1, chap. 4, sec. 6.

9. Hume, *An Enquiry Concerning Human Understanding*, ed. E. Steinberg (Indianapolis: Hackett, 1977), sec. 12, pt. 3, p. 114.

10. Immanuel Kant, *Prolegomena to Any Future Metaphysics*, trans. and intro. L. Beck (New York: The Liberal Arts Press, 1951), Preamble, sec. 2, pp. 14–19; *Critique of Pure Reason*, trans. F. Müller (Garden City: Doubleday, 1966), Introduction, sec. 5, pp. 10–12.

11. Georg Hegel, *Science of Logic*, trans. W. H. Johnston and L. G. Struthers, 2 vols. (New York: Macmillan, 1929), 1:95.

12. Hegel, *Reason in History: A General Introduction to the Philosophy of History*, trans. Robert Hartman (New York: The Liberal Arts Press, 1953), p. 87.

13. Wilhelm Dilthey, *Pattern and Meaning in History*, trans. and ed. H. P. Rickman (New York: Harper and Row, 1961), p. 67.

14. See Ranke's letter to his brother, Ernst Ranke, September 5, 1876, in Leopold von Ranke, *Das Briefwerk*, ed. Walter Fuchs (Hamburg: Hoffmann, Hoffmann und Campe Verlag, 1949), p. 537. His comments on Kant's epistemology are found in a fragment from the 1830s. See Leopold von Ranke, *Tagebücher*, ed. Walter Fuchs (Munich and Vienna: R. Oldenbourg Verlag, 1964), p. 159.

15. For another perspective on Ranke's epistemology, see Hayden White, *Metahistory: The Historical Imagination in Nineteenth-Century Europe* (Baltimore: Johns Hopkins University Press, 1973), pp. 165–67.

16. Leopold von Ranke, *Die römischen Päpste*, 3 vols. (Leipzig, 1867), 1:46.

17. Albert Einstein, *Essays in Science* (New York: Philosophical Library, 1934), pp. 55–56.

18. Bertrand Russell, *The ABC of Relativity*, rev. ed., F. Pirani, ed. (New York: New American Library, 1958), pp. 16–24.

19. Alfred North Whitehead, *Science in the Modern World* (New York: Macmillan, 1925; rpt. 1967), pp. 58–64. This is Whitehead's most accessible writing on the subject. For a more technical discussion, see *Process and Reality* (New York: Macmillan, 1929; rpt. 1969).

20. Sigmund Freud, *The Unconscious* (1915); in *Complete Psychological Works*, trans. and ed. J. Strachey, 24 vols. (London: Hogarth Press, 1957), 14:171.

21. Freud, *Introductory Lectures on Psychoanalysis* (1917), lec. 1; *Complete Works*, 15:22.

22. Freud, *Beyond the Pleasure Principle* (1919), chap. 2; *Complete Works*, 18:12–17.

23. Ernest Jones, *The Life and Work of Sigmund Freud* (New York: Basic Books, 1953), p. 41.

24. Freud, *Civilization and Its Discontents* (1930), chap. 6; *Complete Works*, 21:122.

25. No one has understood this aspect of Freud's thought better than Norman O. Brown, *Life Against Death: The Psychoanalytical Meaning of History* (Middletown, Conn.: Wesleyan University Press, 1959).

26. Gary Zukav, *The Dancing Wu Li Masters* (New York: Bantam Books, 1980), pp. 282–304.

CHAPTER THREE

1. François Chatelet, "Le temps de l'histoire et l'évolution de la fonction historienne," *Journal de psychologie* 53 (1956):356.

2. Virginia Hunter, *Past and Process in Herodotus and Thucydides* (Princeton: Princeton University Press, 1982), p. 167.

3. Herodotus, *Histories*, trans. A. D. Godley, 4 vols. (Cambridge, Mass.: Harvard University Press, 1920–30), bk. 1, chap. 4.

4. R. Ball, "Generational Dating in Herodotus," *Classical Quarterly*, N. S., 29 (1979):276–81.

5. "Khrēn gar Kandaulē genēsthai kakōs." Herodotus, 1.8.

6. As when Darius' soldiers grew weary after a year and seven months before the walls of Babylon, and the king had to adopt a bizarre stratagem for taking the city. Herodotus, 3.152–59.

7. Raphael Sealey, *A History of the Greek City-States ca. 700–338 B.C.* (Berkeley: University of California Press, 1976), chap. 5.

8. H. T. Wallinga, "The Structure of Herodotus II, 99–142," *Memosyne* 4 (1959):204–33.

9. Herman Fränkel, *Wege und Formen früh-griechischen Denkens: literarische und philosophiegeschichtliche Studien*, ed. F. Tietze (Munich: C. H. Beck'sche Verlagsbuchhandlung, 1955), p. 85. Henry Wood, *The Histories of Herodotus: An Analysis of the Formal Structure* (The Hague: Mouton, 1972), p. 12, makes a similar observation about the flexibility of Herodotus' chronology.

10. Thucydides, *History of the Peloponnesian War*, trans. Charles Smith, 4 vols. (Cambridge, Mass.: Harvard University Press, 1919–23), bk. 5, chap. 20.

11. For alternative datings of the period before the war, see A. W. Gomme, *A Historical Commentary on Thucydides*, 5 vols. (Oxford: The Clarendon Press, 1945–81), and Philip Deane, *Thucydides' Dates, 465–431 B.C.* (Don Mills, Ontario: Longman, 1972).

12. For a full discussion of the dating issue, see Deane, pp. 22–38.

13. See the article on Smerdis in Paulys-Wissowa, *Real-Encyclopädie des classischen Altertumswissenschaft* (Stuttgart: Metzler, 1894–1972), R. 2, 5:711.

14. "Dēla dē polloisi tekmērioisi esti ta theia tōn prēgmatōn" (9.100).

15. Virginia Hunter, *Thucydides the Artful Reporter* (Toronto: Hakkert, 1973), pp. 85–94, has discussed this passage in detail.

CHAPTER FOUR

1. Polybius, *The Histories*, trans. W. R. Paton, 6 vols. (New York: G. P. Putnam's Sons, 1922–27), bk. 1, chap. 4.

2. G. W. Trompf, *The Idea of Historical Recurrence in Western Thought: From Antiquity to the Reformation* (Berkeley: University of California Press, 1979), pp. 4–115.

3. Eduard Schwartz, "Die Königslisten des Eratosthenes und Kastor mit Excursen über die Interpretationen bei Africanus und Eusebios," *Abhandlungen der Königlichen Gesellschaft der Wissenschaften zu Göttingen*, 40 (1894–95):1–96, discusses the influence of Eratosthenes' chronology on later chronologists.

4. Polybius, 1.41, where he noted that it was the fourteenth year of the war.

5. "Their successors, Gnaeus Servilius and Gaius Sempronius, put to sea with their whole fleet as soon as it was summer." Polybius, 1.39.

6. Paul Pedich, *La méthode historique de Polybe* (Paris: Les Belles Lettres, 1964).

7. See F. Walbank, *Polybius* (Berkeley: University of California Press, 1972), pp. 105–7, and R. M. Errington, "The Chronology of Polybius' Histories, Books i and ii," *Journal of Roman Studies* 57 (1967):96–108.

8. Bickerman, *Chronology*, p. 77.

9. Livy, *Ab urbe condita*, trans. B. O. Foster, 14 vols. (Cambridge, Mass.: Harvard University Press, 1919–1959), bk. 4, chap. 7.

10. Tacitus, *The Histories*, trans. Clifford Moore, and *The Annals*, trans. John Jackson, 4 vols. (Cambridge, Mass.: Harvard University Press, 1937–51), bk. 11, chap. 11.

11. See T. J. Luce, *Livy: The Composition of His History* (Princeton: Princeton

University Press, 1977), for a thorough analysis of Livy's organization and method. For a more traditional account, see P. G. Walsh, *Livy: His Historical Aims and Methods* (Cambridge: Cambridge University Press, 1961), and Gunter Wille, *Der Aufbau des Livianischen Geschichtswerks* (Amsterdam: Grüner, 1973).

12. See Luce, *Livy*, p. 59. See also F. W. Walbank, *A Historical Commentary on Polybius*, 3 vols. (Oxford: The Clarendon Press, 1957–79), 1:257, for a discussion of how Livy reworked Polybius to identify events happening in the autumn, for which Polybius had no word.

13. Arnaldo Momigliano, "Time in Ancient Historiography," *History and the Concept of Time: History and Theory, Studies in the Philosophy of History* 5, Beiheft 6 (1966):1–23.

14. Josephus, *The Jewish War*, trans. G. A. Williamson (Harmondsworth: Penguin, 1959), p. 325.

15. Eusebius Pamphili, *Die Chronik des Hieronymous*, ed. R. Helm, vol. 7, i and ii, of *Eusebius Werke*, 8 vols. (Leipzig: J. C. Hinrichs, 1902–56).

16. Alden Mosshammer, *The Chronicle of Eusebius and Greek Chronographic Tradition* (Lewisburg: Bucknell University Press, 1979), p. 67.

17. The original arrangement is difficult to recover, as there are important variations in extant manuscripts. See Mosshammer, pp. 38–48.

18. *Praeparatio Evangelica*, 10.9, 1–11, in *Eusebius Werke*, vol. 8, summarized in Mosshammer, pp. 32–34.

19. Erich Caspar, *Die älteste römische Bischofsliste; kritische Studien zum Formproblem des Eusebianischen Kanons* (Berlin: Deutsche Verlagsgesellschaft, 1926). His argument is discussed by Mosshammer, pp. 63–65.

20. He mentioned Eratosthenes in noting a discrepancy in Homer.

21. Eusebius, *Kirchengeschichte*, in *Eusebius Werke*, vol. 2, i and ii; translation from *History of the Church*, trans. G. A. Williamson (Baltimore: Penguin, 1965), bk. 1, chap. 1.

22. Eusebius, *History of the Church*, 3.21, but at 4.1 he contented himself simply with locating the death of the bishop of Alexandria with reference to the reign of Trajan, and said that the bishop of Rome had died "meanwhile."

23. See also 2.11, where he used the famine of Claudius' reign to substantiate an account in Acts.

24. Otto Seel has drawn this conclusion from an impressive analysis of the internal form of Justin's epitome in comparison with the fragments of Pompeius' work which have survived as quotations in other writings. *Eine römische Weltgeschichte. Studien zum Text der Epitome des Justinus und zur Historik der Pompeius Trogus* (Nuremberg: Verlag Hans Carl, 1972). See p. 246 for a discussion of Trogus' chronology.

25. M. *Juniani Justini Epitoma historiarum Philippicarum Pompei Trogi*, ed. Otto Seel, after Francis Ruehl (Leipzig: B. G. Teubner, 1935), bk. 1, chap. 1.

26. Seel, *Eine römische Weltgeschichte*, p. 247.

27. P. Annius Florus, *Epitomae libri ii*, ed. Otto Rossbach (Leipzig: B. G. Teubner, 1896), bk. 1, chap. 1.

28. For a discussion of the issue, see Bessie Walker, *The Annals of Tacitus* (Manchester: Manchester University Press, 1952).

29. Suetonius, *Lives of the Caesars*, trans. J. C. Rolfe, 2 vols. (Cambridge, Mass.: Harvard University Press, 1913; rev. 1951), bk. 3, chap. 18,11–13,29.

30. Dionysius of Halicarnassus, Letter to Pompeius, in *Dionysius of Halicarnassus, The Three Literary Letters*, trans. W. Rhys Roberts (Cambridge: The University Press, 1901), pp. 111–13.

CHAPTER FIVE

1. Peter Brown, *Augustine of Hippo* (Berkeley: University of California Press, 1969), offers an excellent biography.

2. Augustine, *Confessionum libri xiii*, ed. M. Skutella; rev. H. Jürgens and W. Schaub (Stuttgart: B.G. Teubner, 1969), bk. 11, chap. 23; translation from *Confessions*, trans. Rex Warner (New York: The New American Library, 1963).

3. Quoted in Simplicius of Silicia, *In Aristoteles physicorum libros octos commentaria*, ed. H. Diels, 2 vols. (Berlin, 1882–95), 1:789–90.

4. See Paul Ricoeur, *Time and Narrative*, 1:6–30, for another perspective on this tension.

5. Augustine, *De civitate Dei*, ed. B. Dombart and A. Kalb, 2 vols., in Corpus Christianorum, Serie Latina, 47–48 (Tournai: Typographi Brepols, 1955), bk. 22, chap. 30; translation from *City of God*, trans. Marcus Dods et al. (New York: Modern Library, 1950).

6. Augustine, *City of God*, 16.43. Though he left the personal metaphor undeveloped here, elsewhere he elaborated it. See *De Genesi contra Manichaeos*.

7. See Trompf, *Historical Recurrence*, pp. 200–220, for a detailed study of the roots and subsequent development of Age theory.

8. *Historiae Francorum*, in *Gregorii Turonensis opera*, ed. W. Arndt and B. Krusch, 2 vols., in Monumenta Germaniae historica: Scriptores rerum Merovingicarum, 1 (Hannover, 1885), Preface; translation from *Gregory of Tours: The History of the Franks*, trans. Lewis Thorpe (Harmondsworth, Penguin, 1974).

9. Paulus Orosius, *Historiarum adversus paganos libri vii*, ed. Karl Zangemester, in Corpus scriptorum ecclesiasticorum Latinorum, 5 (Vienna, 1882), Dedication to St. Augustine; translation from *Seven Books of History against the Pagans*, trans. Irving W. Raymond (New York: Columbia University Press, 1936).

10. The story of the Easter calendar is complex and is told differently by different scholars. Two particularly clear accounts are found in Reginald Lane Poole, *Chronicles and Annals* (Oxford: The Clarendon Press, 1926), and in *Bedae opera de temporibus*, ed. Charles Jones (Cambridge, Mass.: The Medieval Academy of America, 1943).

11. Prosper of Aquitaine had also made such a cycle in 433 in his abridgment of Eusebius' *Chronicle*. See Poole, *Chronicles*, p. 17.

12. Tacitus, *Annals*, 11.11. His detailed explanation of these calculations in the *Histories* has been lost.

13. Gustav Teres, "Time Computations and Dionysius Exiguus," *Journal for the History of Astronomy* 15 (1984):177–88.

14. Gordon Moyer, "The Gregorian Calendar," *Scientific American* 246 (May 1982):144–53.

15. Bede, *De temporibus*, 1.

16. The critical editions are those cited here. For scholarly studies, see T. Mörner, *De Orosii vita eiusque historiarum libris septem adversus paganos* (Berlin, 1844); Gabriel Monod, *Etudes critiques sur les sources de l'histoire mérovingienne, 1ère partie, Introduction, Gregoire de Tours, Marius d'Avenches* (Paris: A. Franck, 1872); and B. Huber, *Otto von Freising, sein Charakter, seine Weltanschauung, sein Verhältniss zu seiner Zeit und seinen Zeitgenossen als ihr Geschichtschreiber* (Munich, 1847).

17. See Lucien Febvre, *Le problème de l'incroyance au xvie siècle* (Paris: Editions Albin Michel, 1942; 3d ed., 1969), pp. 423–26. See also George Ifrah, *From One to Zero: A Universal History of Numbers*, trans. Lowell Bair, (New York: Viking, 1985).

18. There are indeed copyists' errors. Krusch's edition in the *Monumenta* gives four variants of the last figure in the series (147, 218, 118, 168) and two in the total (5,742 and 5,819).

19. Otto of Freising, *Chronica sive historia de duabus civitatibus*, ed. A. Hofmeister, in Scriptores rerum Germanicarum in usum scholarum (Hannover: Hahnsche Buchhandlung, 1912), bk. 1, chaps. 3,4,5.

20. Orosius, 7.11, where he changed Eutropius' date for Nerva.

21. Bede, *Historiam ecclesiasticam gentis Anglorum*, ed. C. Plummer (Oxford: The Clarendon Press, 1946), 3.27;5.23.

22. Bede, 3.1; translation from *History of the English Church and People*, trans. L. Sherley-Price; rev. R. E. Latham (Harmondsworth: Penguin, 1955; rev. 1968).

23. Robert of Clari, *The Conquest of Constantinople*, trans. Edgar McNeal (New York: W. W. Norton, 1969), p. 31.

24. Otto, 1.5; translation from *The Two Cities*, trans. Charles Mierow, ed. Austin Evans and Charles Knapp (New York: Columbia University Press, 1928).

25. Actually Otto never dated Lothar's coronation and his calculation of the years here is erroneous.

26. He described his own work that way in the dedication, and used the term to refer to Orosius. Otto, 4.21.

27. *Gesta Fridirici I imperatoris*, ed. G. Waitz, in Scriptores rerum Germanicarum in usum scholarum, 46, 2d ed. (Hannover: Hahnsche Buchhandlung, 1884), 2.1.

28. Donald Wilcox, "The Sense of Time in Western Historical Narratives from Eusebius to Machiavelli," in *Classical Rhetoric and Medieval Historiography*, ed. Ernst Breisach, Studies in Medieval Culture, no. 19, Medieval Institute Publications (Kalamazoo: Western Michigan University Press, 1985), pp. 206–9.

29. For a full discussion of medieval treatment of personality in historical narratives, see Wilcox, "Sense of Time," pp. 200–15.

CHAPTER SIX

1. *Le familiari*, ed. V. Rossi and V. Bosco, 4 vols. (Florence: Sansoni, 1933–42), vol. 4, bk. 24, letter 3.

2. Amos Funkenstein, *Heilsplan und naturliche Entwicklung: Gegenwartsbestimmung im Geschichtsdenken des Mittelalters* (Munich: Nymphenburger Verlagshandlung, 1965).

3. See also Theodor Momsen, "Petrarch's Conception of the 'Dark Ages'," *Speculum* 17 (1942):226–42.

4. Petrarch, *Apologia contra eum qui maledixit Italia*, in Petrarca, *Prose*, ed. G. Martelotti et al. (Milan: Riccardo Ricciardi, 1955), p. 790.

5. Salutati, *Epistolae*, May 5, 1379, in E. Emerton, *Humanism and Tyranny: Studies in the Italian Trecento* (Cambridge, Mass.: Harvard University Press, 1925), pp. 300–304. Trompf, *Historical Recurrence*, p. 220, notes that this interpretation reinforces the cyclical aspects of time, but within the traditional Great Week model.

6. Hans Baron, *Crisis of the Early Italian Renaissance* (Princeton: Princeton University Press, 1966).

7. Dante, *On World Government or De Monarchia*, trans. Herbert W. Schneider (New York: The Liberal Arts Press, 1949), 1.16.

8. Salutati, *Invectiva lini Salutati in Antonium Luschum vicentinum*, ed. D. Moreni (Florence, 1826), pp. 28–29.

9. Giovanni Villani, *Cronica*, 4 vols. (Florence, 1844–45), bk. 1, chap. 26, on dating the founding of Rome, and 2.7, on the date of the Lombard invasion.

10. Leonardo Bruni, *Historiarum Florentini populi libri xii*, vol. 19, pt. 3, of Rerum Italicarum scriptores, ed. E. Santini (Città di Castello: Casa Editrice S. Lapi, 1914), bk. 1, p. 19.

11. Flavio Biondo, *Historiarum Romanarum decades tres* (Venice, 1483), bk. 1, chap. 1.

12. See 2.1, where he summarized the significance of the events in the first part of his work.

13. Bruni explained the simultaneous deaths of Castruccio Castracani and Galeazzo Visconti in these terms. Bruni, 5.137.

14. J. G. A. Pocock, *The Machiavellian Moment: Florentine Political Thought and the Atlantic Republican Tradition* (Princeton: Princeton University Press, 1975).

15. See Hans Baron, *Crisis*.

16. See Donald Wilcox, *The Development of Florentine Humanist Historiography during the Fifteenth Century* (Cambridge, Mass.: Harvard University Press, 1969), chap. 4, for a detailed discussion of the argument that follows.

17. See Sallust, *Bellum Jugurthum*, trans. J. C. Rolfe (Cambridge, Mass.: Harvard University Press, 1921; rev. ed. 1931), 52–55, where he discussed long-term changes in strategy after a defeat, returning abruptly to the point where the news of the battle reached Rome.

18. Wilcox, *Development*, pp. 154–65.

19. In 1393, for instance. Bruni, 11.264–65.

20. See Poggio Bracciolini, *Historia Fiorentina*, ed. J. Recanati (Venice, 1715), bk. 3, p. 122, for his treatment of 1393.

21. Michel Foucault, *The Order of Things: An Archaeology of the Human Sciences*, a trans. of *Les Mots et les choses* (New York: Vintage Books, 1973), pp. 50–58.

22. Francesco Guicciardini, *Storia d'Italia*, ed. Silvana Seidel Menchi, 3 vols. (Turin: Giulio Einaudi, 1971), bk. 1. chap. 9.

23. Machiavelli, *Discorsi sopra la prima deca di Tito Livio*, in vol. 1 of *Tutte le opere di Niccolò Machiavelli*, ed. F. Flora and C. Cordié, 2 vols. (Milan: Arnaldo Mondadori, 1949–50), bk. 1, Introduction; translation from *The Prince and the Discourses*, trans. C. Detmold, (New York: Modern Library, 1950).

24. Guicciardini, *Ricordi*, ed. R. Spongano (Florence: Sansoni, 1951); translation from *Maxims and Reflections of a Renaissance Statesman*, trans. M. Domandi (New York, Harper and Row, 1965), no. 117.

25. Guicciardini, *Considerazione sui 'Discorsi' del Machiavelli*, in *Opere*, ed. E. L. Scarano, 3 vols. (Turin: Editrice Torinese, 1970), 1:660–61.

26. Paulo Cortesi, *De hominibus doctis*, in P. Villani, *Liber de civitatis Florentiae famosis civibus* (Florence, 1847).

27. Machiavelli, *Istorie fiorentine*, in vol. 2 of *Tutte le opere*, 3.16–17. See Wilcox, "Sense of Time," pp. 195–97.

28. Donald Wilcox, "Guicciardini and the Humanist Historians," *Annali d'Italianistica* 2 (1984):19–33. See also Mark Phillips, *Francesco Guicciardini: The Historian's Craft* (Toronto: University of Toronto Press, 1977), pp. 157–73, for a discussion of other personalities in Guicciardini's work.

29. See William J. Bouwsma, "The Two Faces of Humanism: Stoicism and Augustinianism in Renaissance Thought," *Itinerarium Italicum: The Profile of the Italian Renaissance in the Mirror of its European Transformations, Dedicated to Paul Oscar Kristeller on the Occasion of his 70th Birthday*, ed. Heiko Oberman with Thomas Brady (Leiden: E. J. Brill, 1975), pp. 3–60, discusses the Augustinian strand in Renaissance humanism.

30. He is more likely to mention the jubilee when it is celebrated early by a greedy pope. See 1.32.

31. Biondo, 1.31, had located the Roman revolt "in the same year"; Machiavelli simply says "in this time."

CHAPTER SEVEN

1. Voltaire, *The Age of Louis XIV*, trans. M. Pollack, (London: J. M. Dent, 1926), p. 1.

2. Edward Gibbon, *The History of the Decline and Fall of the Roman Empire*, ed. J. B. Bury, 7 vols. (London: Methuen, 1896), 1:78.

3. Donald Kelley, *The Foundations of Modern Historical Scholarship* (New York: Columbia University Press, 1970).

4. Julian Franklin, *Jean Bodin and the Sixteenth Century Revolution in Law and History* (New York: Columbia University Press, 1963).

5. Bodin, *Methodus ad facilem historiarum cognitionem* (Amsterdam, 1650; reprint, Aalen: Scientia Verlag, 1967); translation from *Method for the Easy Comprehension of History*, trans. Beatrice Reynolds (New York: Columbia University Press, 1945; reprint, New York: W. W. Norton, 1969), p. 42.

6. George Huppert, *The Idea of Perfect History: Historical Erudition and Historical Philosophy in Renaissance France* (Urbana: University of Illinois Press, 1970), p. 138.

7. La Popelinière, *Idée de l'histoire accomplie*, quoted in Huppert, p. 141.

8. La Popelinière, p. 85, quoted in Huppert p. 141 n. 18.

9. La Popelinière, *L'Histoire des Histoires* (Paris, 1599), p. 20.

10. Scaliger, *Opus novum de emendatione temporum* (Paris, 1583).

11. Anthony Grafton, "Joseph Scaliger and Historical Chronology: The Rise and Fall of a Discipline," *History and Theory* 14(2) (1975):156–85. Grafton's work is a remarkably clear and thorough analysis of Scaliger's contribution and the following discussion draws heavily on it.

12. Scaliger, p. 3, quoted in Grafton, p. 162.

13. Scaliger, *Isagogici canones*, in vol. 2 of *Thesaurus Temporum* (Lyons, 1606; reprint, Osnabrück: Otto Zeller, 1968), p. 117.

14. "More mathematicorum." Scaliger, *Isagogici*, p. 274.

15. M. St. Allais, ed., *L'art de verifier les dates*, 41 vols. (Paris, 1819–44), 1:1.

16. The basic material for the biography of Petavius is found in François Oudin's article in vol. 37 of *Mémoires pour servir à l'histoire des hommes illustres dans la république des lettres*, ed. J. P. Niceron (Paris, 1737), pp. 81–234. The more recent work of J. C. Vital Chatelain, *Le Père Denis Petau d'Orléans, Jésuite. Sa vie et ses oeuvres* (Paris, 1884), contains little that is not found in the Niceron volume. Pietro di Rosa's "Denis Petau e la cronologia," *Archivium historicum Societatis Jesu*, 29 (1960):3–54, is largely concerned with his chronological method.

17. Petavius, *Opus de doctrina temporum*, ed. J. Hardouin, 3 vols. (Venice, 1705), Letter of Dedication, 1.xxxiii.

18. Petavius, *Rationarium temporum* (Paris, 1633), Preface.

19. G. J. Vossius, *De theologia gentili et physiologia Christiana. . . .* , 3d ed. (Frankfurt, 1675), 28:212, cited in Grafton, "Scaliger," pp. 175–76.

20. Morus, *Thesaurus temporum*, Preface, cited in Niceron, *Memoires*, pp. 111–12.

21. Manuel, *Isaac Newton*, p. 266 n. 3.

22. Newton, *The Chronology of Ancient Kingdoms Amended*, in vol. 5 of *Isaaci Newtoni opera*, ed. S. Horsley (London, 1785), p. 185.

23. *Bibliotheke Historike*, 1.51.

24. From a New College manuscript published in Manuel, *Isaac Newton*, p. 195.

25. New College manuscript, quoted by Manuel, p. 66.

26. Though his reasons behind the change in the date of the Argonautic expe-

dition are somewhat obscure in the published *Chronology*, Frank Manuel has found in the manuscripts in New College a more detailed account of his reasoning. What follows is a brief summary of that account. For a more thorough analysis see Manuel, chaps. 4 and 5.

27. New College manuscript, quoted by Manuel, p. 276, n. 21.

28. Richard S. Westfall, "Newton and the Fudge Factor," *Science* 179 (1973):751–58.

29. Jean Hardouin, "Le Fondement de la chronologie de M. Newton, Anglois, imprimé à Londres en 1726, sappé le P.J.H," *Mémoires de Trévoux* 29 (1729; reprint, Geneva: Slatkine Reprints, 1968), 1570–77.

30. V. Grummel, *La chronologie*, in vol. 1 of *Traité d'études Byzantine* (Paris: Presses universitaires de France, 1958), p. 56.

31. Gibbon, *Memoirs of my Life*, ed. G. A. Bonnard (London, 1966), quoted in Grafton, "Scaliger," p. 156.

32. Giovanni Battista Vico, *Principi di scienza nuova*, in *Opere*, ed. Fausto Nicolini (Milan: Riccardo Ricciardi, 1953), par. 740; translation from *The New Science of Giambattista Vico*, trans. T. Bergin and M. Fisch (Ithaca: Cornell University Press, 1968).

33. *Dell'antichissima sapienza Italica*, in *Opere*, p. 254.

34. See A. Robert Caponigri, *Time and Idea* (London: Routledge and Kegan Paul, 1953).

CHAPTER EIGHT

1. Aristotle, *Poetics*, 9; translation from *The Basic Works of Aristotle*, ed. and intro. Richard McKeon (New York: Random House, 1941).

2. Thucydides, 1.97;5.20. Lowell Edmonds, *Chance and Intelligence in Thucydides* (Cambridge, Mass.: Harvard University Press, 1975), p. 155, discusses his use of the term.

3. Christoph Schneider, *Information und Absicht bei Thukydides: Untersuchung zur Motivation des Handelns* (Göttingen: Vandenhoeck und Ruprecht, 1974).

4. H. L. Hudson Williams, "Conventional Forms of Debate and the Melian Dialogue," *American Journal of Philology* 71 (1950):156–59.

5. A. Andrewes, "The Melian Dialogue and Perikles' Last Speech," *Proceedings of the Cambridge Philological Society* 186 (1960):1–10.

6. The accuracy of Thucydides' speeches is not the central concern of scholars that it once was. Though F. E. Adcock devoted a whole chapter of his 1963 book to the subject of where Thucydides could have gotten accurate accounts of the speeches (*Thucydides and His History* [Cambridge: Cambridge University Press, 1963]), a collection of essays devoted to the speeches in 1973 produced many important insights into their function and no essays directly concerned with assessing their factual accuracy. Philip Stadter, ed. *The Speeches in Thucydides* (Chapel Hill, N. C.: University of North Carolina Press, 1973). Still more recently Marc Cogan's *The Human Thing: The Speeches and Principles of Thucydides' History* (Chi-

cago: University of Chicago Press, 1981) is principally concerned with function and interpretation.

7. *Frutolfs und Ekkehards Chroniken und die anonyme Kaiserchronik*, ed. F. J. Schmale and I. Schmale-Ott (Darmstadt: Wissenschaftliche Buchgesellschaft, 1972), 182.13.

8. Lorenzo Valla, *Historiarum Ferdinandi Regis Aragoniae libri III*, in *Opera omnia*, ed. E. Garin, 2 vols. (Turin: Bottega d'Erasmo, 1962), 2:29. Linda Gardiner Janik has treated this debate in "Lorenzo Valla: The Primacy of Rhetoric and the De-moralization of History," *History and Theory* 12 (1973):389–404.

9. Bartolommeo Fazio, *Invective IIII*, in *Le invettivi di Bartolommeo Fazio contro Lorenzo Valla*, ed. Roberto Valentini, Licei Rendiconti XV (1906):493–550, p. 527.

10. Valla, *Recriminationum in Bartolomaeum Facium libri IIII*, in *Opera*, 1:574, quoted in Janik, p. 399.

11. Valla, 1.591, on the public records of Catalonia, cited in Janik, p. 403.

12. Valla, *De falso credita et ementita Constantini donatione declamatio*, in *Opera*, 1:765–66,765,775–76.

13. Silvana Seidel Menchi in her edition of the *Storia d'Italia* has documented Guicciardini's method clearly.

14. See Wilcox, "Sense of Time."

15. Domenico Buoninsegni, *Historia fiorentina* (Florence, 1581), pp. 594–95.

16. Machiavelli, *Discourses*, 2.1; translation here and in the next two excerpts is mine.

CHAPTER NINE

1. See Hans Peter Stahl, *Thukydides: Die Stellung des Menschen im geschichtlichen Prozess* (Munich: Beck, 1966), chap. 1.

2. Marcel Proust, *Remembrance of Things Past*, trans. C. K. Scott Moncrieff et al., 3 vols. (New York: Random House, 1981), 3:555.

3. Erik Erikson, *Young Man Luther: A Study in Psychoanalysis and History* (New York: W.W. Norton, 1962), p. 251.

4. Michel Foucault, *The Order of Things: An Archaeology of the Human Sciences* (New York: Vintage Books, 1973), p. 340. Foucault's contributions in this area have not always been well understood by his readers. See Pamela Major-Poetzl, *Michael Foucault's Archaeology of Western Culture: Toward a New Science of History* (Chapel Hill: University of North Carolina Press, 1979), for an excellent analysis of the affinities between Foucault's concepts of time and space and those used in modern physics.

5. Jacques Le Goff, "Is Politics Still the Backbone of History?" *Daedalus* 100 (1971):1–13.

Index

Abraham, 105, 127, 169

Absolute history: dating of, 196–200; origins of, 189–90; scope of, 194, 202–3

Absolute subject, 23–31, 269; relative time and, 43; empiricism and, 30; Freudian critique of, 44–45; Kant and, 33–34

Accuracy. *See* Veracity

Absolute time: astronomy and, 22–23, 203, 209; defined, 2–3, 16; historical truth and, 38, 243–44; identity and, 255–56; Newton's chronology and, 23

Achaean League, 88

Adcock, F. E., 283n

Acton, Lord, 5, 7

Africanus, Sextus Julius, 105

Akribeia, 221, 224; defined, 225; difference from alētheia, 228–29. *See also* Veracity

Alētheia, 221, 224; defined, 225; difference from akribeia, 228–29; narrative and, 229; Polybius and, 229–30; speeches and, 225–28. *See also* Versimilitude

Amasis, 60

Amenophis, 210, 212

Andrewes, A., 283n

Annales School, 10, 271

Annalistic form, 93–104, 166–72; Bruni's use of, 167–71; Guicciardini's use of, 182–85; linear time and, 95; Livy's use of, 95–97; order of events in, 97–98; personal time and, 116–17; Polybius' use of, 92; relative time and, 94–95; Tacitus' use of, 97–98, 101–3

Anno Domini, 129–36, 137, 143–44, 149

Apollinaire, Guillaume, 3

Apollodorus of Athens, 104

Archaeology: in Foucault, 269–70; in Herodotus, 53–54; in Machiavelli, 175–77; in Thucydides, 69–74, 233–34, 270; in Vico, 216–19

Archons, dating by, 59

Argonautic expedition, 210–12

Aristotle, 155; on history, 222; time and, 120–21

Arrian, 123

Athens, character of, 68–69

Atticus, 153

Auerbach, Erich, 14

Augustine, 119–29, 139, 144, 155, 177, 193; astronomical time and, 120–21; career, 119–20; chronology and, 122–24, 128; *City of God,*

Index